Textual Healing:
Studies in Medieval English Medical, Scientific and Technical Texts

Linguistic Insights

Studies in Language and Communication

Edited by Maurizio Gotti,
University of Bergamo

Volume 101

PETER LANG
Bern · Berlin · Bruxelles · Frankfurt am Main · New York · Oxford · Wien

Javier E. Díaz Vera & Rosario Caballero (eds)

Textual Healing:
Studies in Medieval English Medical, Scientific and Technical Texts

PETER LANG
Bern · Berlin · Bruxelles · Frankfurt am Main · New York · Oxford · Wien

Bibliographic information published by Die Deutsche Nationalbibliothek
Die Deutsche Nationalbibliothek lists this publication in the Deutsche Nationalbibliografie; detailed bibliographic data is available on the Internet at ‹http://dnb.d-nb.de›.

British Library and Library of Congress Cataloguing-in-Publication Data:
A catalogue record for this book is available from *The British Library*, Great Britain.

Library of Congress Cataloging-in-Publication Data

Textual healing : studies in medieval English medical, scientific and technical texts / Javier E. Díaz Vera & Rosario Caballero (eds).
p. cm. – (Linguistic insights: studies in language and communication; v.101)
Includes bibliographical references and index.
ISBN 978-3-03911-822-9 (alk. paper)
1. English language–Middle English, 1100-1500–Medical English. 2. Medicine, Medieval–England. 3. English language–Middle English, 1100-1500–Technical English. 4. Scientific literature–England–History–To 1500. 5. Medical literature–England–History–To 1500. 6. England–Intellectual life–1066-1485.
I. Díaz Vera, Javier E. II. Caballero, Rosario, 1963-
PE664.3.T43 2009
820.9'3561–dc22

2010011828

ISSN 1424-8689
ISBN 978-3-03911-822-9

Hochfeldstrasse 32, CH-3012 Bern, Switzerland
info@peterlang.com, www.peterlang.com, www.peterlang.net

Printed in Germany

Contents

Introduction

More than ten years ago, David Crystal (1997: 106) stated that "most of the scientific, technological and academic information in the world is expressed in English and over 80% of all the information stored in electronic retrieval systems is in English". In the last decades, the use of English has in fact extended greatly to most scientific and technical fields. However, it is seldom recognised that the first steps towards the vernacularisation of scientific and technical writing took place already in the Anglo-Saxon period, which implies that nowadays English is the living language with a longer tradition in specialised writing.

As the growing body of research published in recent years has shown, analysing the language of specialised texts is an opportunity to obtain access to the early history and vernacularisation of learned written styles. Moreover, we need to know more about earlier stages of language use in order to understand the characteristics of present-day specialised writings.

This volume comprises a number of significant contributions to the field of specialised writing in Old and Middle English. The papers collected in this volume offer a wide number of specific interests and approaches to the historical analysis of English specialised texts. This is an area of study in which all the contributors have considerable expertise, which affords them to present data findings while discussing important methodological issues. In addition, in most cases data derive from specially-designed second-generation corpora, reflecting state-of-the-art approaches to historical linguistics, discourse analysis and pragmatics.

In "Null Objects in Middle English Medical Texts" (pp. 1-27), Francisco Alonso-Almeida proposes an approach to null objects (i.e. zero complementation) in Middle English. The interest of his work lies on several factors. In the first place, the study has been conducted using corpus tools and techniques (the Middle English Medical Texts or MEMT corpus, thus basing the results on real data rather than on isolated examples. In the second place, the author contextualises his research by using real texts and, more specifically, various genres from the realm of medicine in order to see the motivations underlying the phenomenon under survey. By so doing, the author points to the role of context and culture in the disambiguation of zero complementation, as well as to the weight of register and genre in the distribution and use of null objects.

Graham D. Caie provides a brief survey of the work done with the Hunterian Collection at the library of the University of Glasgow in his paper "The Hunterian Collection at the University of Glasgow" (pp. 29-34). After introducing Dr William Hunter and providing some interesting facts of his biography and the reasons that led him to donate his collection of old manuscripts to the University of Glasgow, Caie focuses on some of the medical manuscripts donated, describing them in some detail.

In his paper "Line-final Word Division in Late Middle English *Fachprosa*: G.U.L. MS Hunter 497 (V.7.24)" (pp. 35-53), Javier Calle-Martin deals with word division in Middle English manuscripts –a phenomenon often attributed to (a) the apparently erratic practice of mediaeval scribes and, therefore, disregarded, and (b) the 'rule' described by some authors that only two completing letters could be carried over to the second line of the text. Drawing upon research that looks into some of the morphological and/or phonological rules underlying the aforementioned practice, in this paper Calle-Martin attempts to characterize the concept of syllable boundary for mediaeval scribes by classifying the type of splitting and offering a taxonomy of the main word-division rules and, by so doing, shed new light on the phenomenon against other classical works.

Mark Chambers ("*What is this, a betell, or a batowe, or a buskin lacyd?*: Lexicological Confusion in Medieval Clothing Culture", pp. 55-75) focuses on the knotty topic of clothing and textiles in medieval allegorical drama. In an attempt to disentangle the confusion that the aforementioned fashion topic usually creates to commentators of medieval history and literature, Chambers looks at several graphic and textual data to provide an explanation of some terms –which goes from tracing their history to describing in some detail their etymology and use in different periods. The paper ends with an overview of the Lexis of Cloth and Clothing Project launched in 2007, whose aim is to create a searchable database of lexical, graphic and archaeological evidence of medieval cloth and clothing and its lexis in several languages (e.g. Anglo-Norman, Cornish, Flemish, Irish, Middle English, Norn, Norse, Old English, etc.). The ultimate goal is to build a useful and polyglot resource for scholars of clothing history, material culture and historical linguistics.

Javier Diaz-Vera's contribution, "Analysing the Diffusion of Scientific Metaphors through a Corpus of Middle English Medical Texts" (pp. 77-95), draws upon the approach to metaphor expounded by George Lakoff to describe the origin, usage and diffusion of medical metaphors as these appear in a variety of genres compiled in the Middle English Medical Texts (MEMT) corpus. Thus, after a brief survey of work done on metaphor in specialised texts, Diaz-Vera explores the metaphors in the surgical texts, specialised texts, and remedies and *materia medica* in the MEMT, sometimes tracing the metaphors back to other types of non-specialised texts. Among his findings, we have the target domain of medicine construed in architectural terms as in THE BODY IS A CASTLE, in textile terms as in THE SKIN IS A GARMENT or in mechanicist terms as in THE BODY IS AN INSTRUMENT. The first metaphor suggests a pervasive use of WAR as the source for medical metaphors such as MEDICINES ARE WEAPONS or SICKNESS IS A CONQUEROR/DEVOURER, all of which illustrate conceptual metaphor as this is described within the Lakoffian tradition.

In her paper "Punctuation Practice in G.U.L. MS Hunter 509" (pp. 97-111), Laura Esteban Segura provides an analysis of the punctuation system found in a Middle English medical manuscript, Glasgow University Library, MS Hunter 509. A description and classification of the use and function of the scribal marks of punctuation is also offered. The resulting data are helpful in order to reach conclusions about the coherence of medieval punctuation practice. When possible, correspondences with modern counterparts are supplied.

María Teresa Marqués-Aguado proposes a dialectogical analysis of the same manuscript in her paper "The Dialectal Provenance of G.U.L. MS Hunter 513 (ff.1r – 37r)" (pp. 113-125). Following Dossena and Lass' suggestion that texts other than literary should be examined from a dialectal perspective, she presents an analysis of the written variants found in the medical treatise under study in an attempt to provide a likely place of provenance.

In "An *Annotated Corpus of Middle English Scientific Prose*: Aims and Features" (pp. 127-145) David Moreno-Olalla and Antonio Miranda-García introduce the *Annotated Corpus of Middle English Scientific Prose* (*ACMESP*), a new annotated corpus that records several late Middle English *Fachprosa* treatises. The aims of the project are not only to develop a corpus, but also to recreate the physical browsing experience of a MS in a virtual environment, yet transcending the merely visual intentions of similar proposals by offering *a*) a physical description of the MSS, to make the visitors' experience all the more rewarding and informative through the coverage of some features that are not immediately evident in virtual environments (for example, measures, quiring or watermarks), and *b*) transcriptions of the texts, for the benefit of those untrained in mediaeval palaeography.

Nadia Obegi-Gallardo's contribution to this volume, "Lexical study of *The Book of Operation* (G.U.L. MS Hunter 95 (T.4.12) ff. 82r -156v): illnesses and treatment" (pp. 147-158), aims at a) classifying the remedies mentioned in *BO*, mostly plant names, into compound or simple nouns and, more specifically, analysing the etymology and permanence of these words through time together with their collocations; b) providing a classification of the illnesses mentioned in the treatise, attending to their survival until Present Day English; c) studying those verbs which help the scribe explain the surgical operations and other ways of healing; and d) explaining the main surgical instruments and treatments which are employed in *BO*.

In her paper "Orality in the Middle English Medical Recipes of G.U.L. Hunter 185" (pp.159-182), Elena Quintana-Toledo explores the degree of orality which may have been transferred from spoken language to a corpus of medical recipes excerpted from the fifteenth-century manuscript G.U.L. Hunter 185, an example of a medieval remedy book. The search for data relies on Ong's characterisation of orally based expression as specified in his seminal work *Orality and Literacy. The Technologizing of the Word* (1982) as it stands as one of the few texts that offers a fairly detailed description of oral style features, though from a macro level viewpoint. Then, she provides a micro level analysis by concentrating on their possible pragmatic interpretation.

Finally, in "Early English Scientific Writing: New Corpora, New Approaches" (pp. 183-212), Irma Taavitsainen shows in detail how corpus-linguistic methods can be combined with qualitative contextual assessments. Her focus is on new developments in corpus compilation, and how the new material collected using these developments can be applied to answer new research questions that deal with the diachronic development of genres and discourse forms and, even more profoundly, with the changing thought-styles of medical writing. The author presents some new insights and demonstrates how a more diversified picture of the developments in the medical register has emerged, with an emphasis on the communicative function of genres in the sociohistorical framework of the period.

The result is a collective volume which illustrates the work of a wide range of scholars in different fields of English Historical Linguistics and their approaches to the analysis and description of medieval English medical and technical writings.

FRANCISCO ALONSO-ALMEIDA

Null Objects in Middle English Medical Texts

1. Introduction

Null objects have been the focus of several studies from both a synchronic and a diachronic standpoint. The diachronic part, however, has not been sufficiently exemplified because the authors's intentions in all the studies consulted (Görlach 1992, Culy 1996, Bender 1999) have been to show the tradition of null objects in the history of English rather than to thoroughly explore their distribution in a fairly well represented corpus of historical texts from past periods of English. In spite of this, these studies have opened the door to more specific diachronic research so that extensively documented descriptions could be done to see whether null complementation had something of an old and well-established existence or its presence in Middle English texts was rare.

This chapter aims to describe null objects in Middle English. For this, I have selected texts from the Middle English Medical Texts (MEMT) corpus. This corpus has proven to be adequate for the study of historical English, as shown by the studies carried out by Taavitsainen (2006), and Alonso-Almeida (forthcoming). My main interest for the study of zero complementation arises from the divergent views in Culy (1996) and Görlach (1992) in their analysis of Middle English recipe texts. My intention is to revise the statements in both works and corroborate one or another with new data and research to elucidate the variety mostly used in the period surveyed. The present investigation considers other textual genres within the medical register that might shed light on the motivations for null objects.

This work is divided into different parts. The next section explores earlier studies that have tried to linguistically describe the internal or external licenses for zero complementation. After that, I introduce the MEMT corpus and the type of data I am interested in for analysis together with a statement on the methodological procedures followed in the present survey. The next sections offer the results and discussion of the findings.

2. Background

One of the earliest studies in the omission of the object in English is Ohlander (1944-1945). The author examines several examples of null complements in the history of English, and he concludes that, in the particular case of coordinated sequences, omissions are a matter of style. Repetitions are stylistically tedious and cumbersome. More interesting in his paper are (a) his view of zero anaphora as a notional strategy to make two parts of discourse become closer, and especially, (b) his cognitive approach to null complements:

> From the point of view of descriptive grammatical analysis, a certain element may therefore be said to be missing or omitted in the speech-unit in question, which may consist of a sentence, clause, word-group, or even a single word. To the average hearer or reader, however, the expression may appear complete enough, since, in most cases, the element to be understood or supplied is so self-evident that the gap is mentally filled in by the audience more or less unreflectingly (105).

In fact, Ohlander touches mostly naively upon many of the issues, which will later form the body of the literature on zero complements. The null-object triggering factors can be grouped into (a) syntactic, (b) pragmatic, and (c) discursive. I will briefly discuss all of them with examples taken from the same literature, or my own.

From a syntactic standpoint, a particular verb may appear with an object or without it, in which case it could be implicit by the action expressed by the semantics of the verb. A sentence like the one *I ate at seven today* may be reanalysed *as I ate my dinner at seven today*. According to Fillmore (1986), the problem with verbs such as *eat* is that we cannot fully predict the understood argument, even if there is any salient argument in the context serving as referent of the elided portion, as in: *My mother prepared shepherd's pie and vegetables for dinner. Mary and I ate there.* There is nothing in the context that could tell us with a high degree of certainty what each of them had for dinner exactly; even if we are tempted to use "shepherd's pie" and "vegetables" as the valid understood arguments. Only the generic "dinner", "meal", "something" or even "stuff" could be used here, although, as pointed out by Groefsema (1995: 144), "it is not clear what these words [something and stuff] themselves represent, or where they come from". Fillmore (1986) calls this type of understood arguments in verbs as Indefinite Null Complements (INC). Verbs showing INC are *eat, drink, read* and *sew*, for instance.

Some other verbs allow the omission of the complement as far as this phonologically non-realised object can be understood and easily retrieved from the context; Fillmore refers to them as Definite Null Complements (DNC). Examples of these verbs are *accept* and *wait.* However, there are some verbs which do not admit the omission of the understood arguments, although the pragmatic context makes the referent clear, as in: *Did you lock?* (Fillmore 1986: 98). The verb *lock* requires the presence of the object even if the speaker knows the door being referred to.

There are two other restrictions regarding DNC. The first one is that semantically related verbs do not present a similar behaviour with respect to their understood arguments, and, whereas *She found out* is fine, *She discovered* is just not possible (Fillmore 1986: 99). The second restriction relates to polysemous verbs. Some verbs like *leave* allow the omission of the argument, but other times the argument must be obligatorily given, as in *She left home* and *She left the package.* The former can be reduced to *She left* without affecting the meaning of the utterance, the second cannot undergo a similar process without a drastic change in the intended meaning (Fillmore 1986: 101). Fillmore (1986: 97) explains this phenomenon arguing that the semantic frames of some verbs have their arguments marked for omission. García Velasco and Portero Muñoz (2002: 8) admit Fillmore's proposal to be "essentially correct" contrary to what other scholars think. They use a different terminology for INC and DNC, being IO (indefinite object) and DO (definite object), respectively.

Junker et al. (2004) see null objects as cases of an isomorphic mismatch between the semantic frames of verbs and their syntactic realisations. Contrary to Fillmore, they claim that the fact that arguments can be omitted means that, semantically, these arguments constitute a variable. This variable "occurs free, in which case the utterance reads as if it had an extra deictic. Sometimes the variable occurs bound in which case the utterance sounds quantificational" (Junker et al. 2004: 17). Verbs like *eat* do not fall within any of these two cases; in their opinion, also recorded in Iten et al. (2006), they are intransitive with some arguments marked for optional addition. I will come to this in section "Further thoughts on null complements" below.

Apart from tackling the problem of the omission feature in transitive verbs, Iten et al. (2006) also comment on the role of pragmatics in licensing null complements in utterances. They base on the work by Groefsema (1995) for this, because they acknowledge she is "on the right track" (Iten et al. 2006: 14), but with some reservations. Their conclusion is that a unified syntactic and pragmatic theory would offer an answer, since "semantics alone cannot explain the difference between verbs that do and verbs that do not allow complements to be omitted" (Iten et al. 2006: 14). Groefsema (1995) pleads for a semantic/pragmatic interpretation of the phenomenon of null objects, but she makes manifest that other properties in the language may also contribute to the licensing of understood arguments. Her proposal is wholly pragmatic in the sense that an argument can be left implicit if there are enough contextual implications as to allow the speaker to effortlessly retrieve the missing information in line with the pragmatic principle of relevance. If there are not enough assumptions to make the speaker meet a satisfactory interpretation, this stands as a constraint to disfavour the presence of an understood argument, at least in the case of English. The semantic contribution of the verb also takes a role in Groefsema's account for they put a selection restriction on the inherent participant so that the interpretation of a given utterance is optimally relevant.

Iten et al. (2006: 11) have taken advantage of Groefsema's self admission that her theory does not seem to work well with other languages to alert on some flaws in her approach:

> If her proposal only works for English verbs, and not for the very same verbs in other languages, then hers just cannot be the right kind of approach – because pragmatics, in the sense she has in mind, is surely universal. And, not to put a fine point on it, her proposal does not work for other languages.

Another point of disagreement has to do with Groefsema's claim that "every verb should allow null complements sometimes" (Iten et al. 2006: 13). For them, this statement is fallacious in the sense that many speakers use elliptical sentences; an utterance such as *Both hands* may be understood by the hearer without spelling out the complete sentences, given the adequate environmental assumptions (Iten et al. 2006: 13). The authors argue that one such context could be a mother ordering her son to use his hands during dinner to hold a glass: *Use both hands or Hold the glass with both hands*. All this stands as an opposition to Groefsema's postulates regarding language and relevance: "it just is not true that it is inevitably more relevant to indicate linguistically, and within the sentence, which item is in question – either via verb with right selection restrictions, or via appropriate words" (Iten et al. 2006: 13).

Culy (1996) opts for an approach based on register, but also on syntax. The author sees arguments as variables x and y in the semantics of a given verb. He mentions that a verb implies at least the following five aspects in the lexicon:

-its valency (i.e. the number of arguments it has),
-the thematic role(s) of each argument,
-the grammatical function of each argument,
-the category of the argument, and
-its meaning (Culy 1996: 13).

He shows the analysis of the verb mix to show that its syntactic representation is similar to its semantic representation:

verb: mix $$\begin{bmatrix}\text{SUBJ}\\ \text{AGT}\\ x\end{bmatrix}, \text{NP}: \begin{bmatrix}\text{OBJ}\\ \text{TH}\\ y\end{bmatrix}$$

subcategorization: NP:
semantics: "mix (x, y)"

He states that, whenever an argument does not appear overtly in the syntax, then that argument is a free variable in the semantics. Free variables are interpreted according to register (I call it genre). He uses the examples of the user's manuals to assert that

> the usage component (the "user manual") specifies the characteristics of registers, and within each register, can specify characteristics of different styles. The specifications may be either about structures that are available to most or all varieties (e.g., zero objects) or about structures that are restricted to a few (or just one) varieties (e.g., *thee* and *thou*) (Culy 1996: 114) [For more recipe genre characteristics, see Taavitsainen 2001 and Carroll 1999].

We will see in the development of this paper that many of the ideas posed by all these scholars are not mutually exclusive. Quite the contrary, they all give clues on how to look at null complements, and, probably, a unified account that reconciles many of the approaches commented on above will give a satisfactory solution to the analysis of this null complement phenomenon.

3. Data and methodology

The texts of this research form a corpus of 129,743 words, collected from the Middle English Medical Texts (MEMT) corpus and analysed using Corpus Presenter software by Hickey (2003), although manual analyses have also been performed at times. The MEMT corpus of texts from 1375 to 1500 has been compiled by Taavitsainen, Pahta and Mäkinen (2005), and it represents only one part of a larger diachronic corpus, which will also include early Modern English material (1500-1700). The texts in this first part are mostly from the fifteenth century or earlier, and they represent varied medical contexts and textual traditions. MEMT is a multigenre compilation in that several types of documents have been collected. Texts are classified into three categories according to their tradition of writing, contents and audience following Voigts (1982, 1984). These three groupings are: surgical texts, specialised texts, and *materia medica*.

Surgical texts are texts belonging to university tradition, and offer highly detailed descriptions. Specialised texts also exemplify the academic tradition and they give information on natural philosophy aspects, including reproduction and medical astrology. The last group, i.e. remedies and *materia medica*, belongs to the remedy tradition and includes recipes, charms and herbals.

The texts for this study have been randomly selected in a distribution of about one third for each tradition:

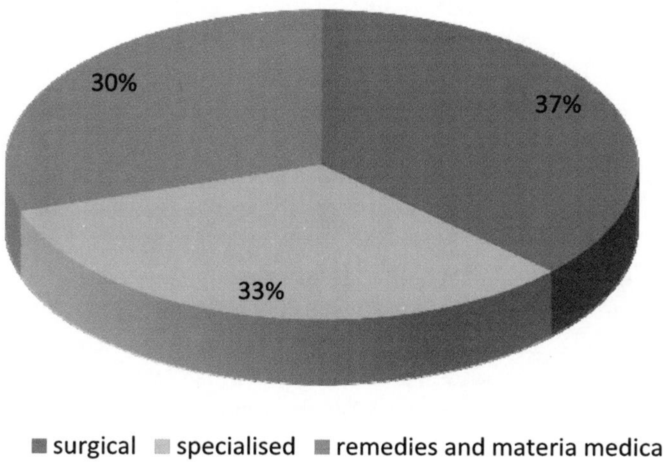

Figure 1: Text traditions in the study.

The type of data I focus on is coordinated verbal phrases by conjunction *and*. This structure gives the context for complementation and for the occurrence of zero anaphors and similar structures. Culy (1996) has argued that this type has not been studied in detail, not even in contemporary English. The phenomenon, then, remains almost unstudied despite the interest of scholars to investigate on the distribution of zero complementation in English.

I have inquired the corpus for the following eight combinations:

1. verb + noun and verb: *take the apple and peel.*
2. verb and verb + noun: *take and peel the apple.*
3. verb + noun and verb + pronoun: *take the apple and peel it.*

4. verb + pronoun and verb + noun: *take it and drink the medicine.*
5. verb + pronoun and verb + pronoun: *take it and drink it.*
6. verb + pronoun and verb: *take it and drink.*
7. verb and verb + pronoun: *take and drink it.*
8. verb + noun and verb + noun: *take the medicine and drink the medicine.*

The examples here given show the type of data I am interested in. Although some scholars have argued that some detransitivised verbs like *eat* should be generally included in the counting of zero anaphors (as in Culy 1996), I think that this very much depends on the context in which this type of verbs are embedded. One case in point is *take the apple and eat* vs. *I ate at twelve.* The first example would count as an instance of zero anaphor for Culy, whereas the second one would not. I do not share this point with Culy for the reasons I will explain later in the following sections.

The instances for analysis have been collected through the inquiry "and" that retrieved hundreds of examples for each text with a context of 30 characters both right and left flanks. Later, manual analyses were performed to discard useless instances, and useful ones were categorised and counted for final statistics. In the following section, I offer the results and further discussion of these findings.

4. Results

The structures found in the texts are, as shown in the pie chart below (samples normalised to 1,000):

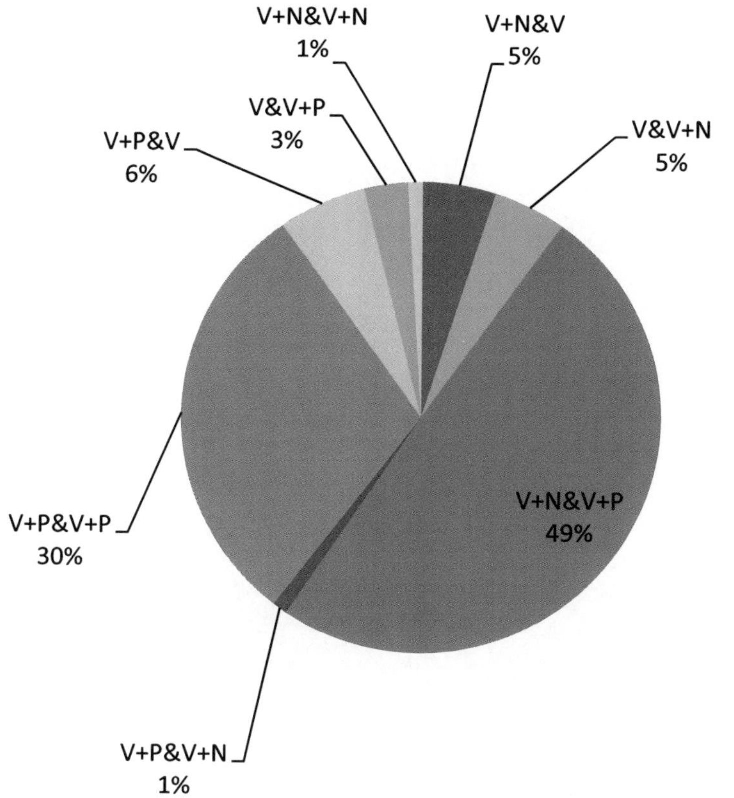

Figure 2: Total occurrences per structure.

The most recurrent syntactic patterns involve the presence either of an overt object (50% of the cases) or of a proform (30% of the cases). Zero anaphora is also significant in that it appears in 14% of the cases, followed by the unexpected, and often ignored, zero cataphora occurring 5% of the instances.

Medical works, Stockholm verse recipes, *wyse book of maystyr peers*, the Gilbert of England compendium and crophill's book present a high frequency of overt complements. Of these, the Stockholm verse recipes and the medical works together with nature of women favour the presence of pronouns as objects. Cataphora presents a more restricted distribution, mainly *properties of things* and *wyse book of maystyr peers.* Anaphora appears in Stockholm verse recipes and *properties of things* with the explicit object being a noun in the coordinated sequence, whereas the *wyse book of maystyr peers* prefers a pronoun instead in the same syntactic context. Remedies seem to show instances of all structure-types in different distribution.

The following three graphs illustrate the distribution of frequencies per textual tradition:

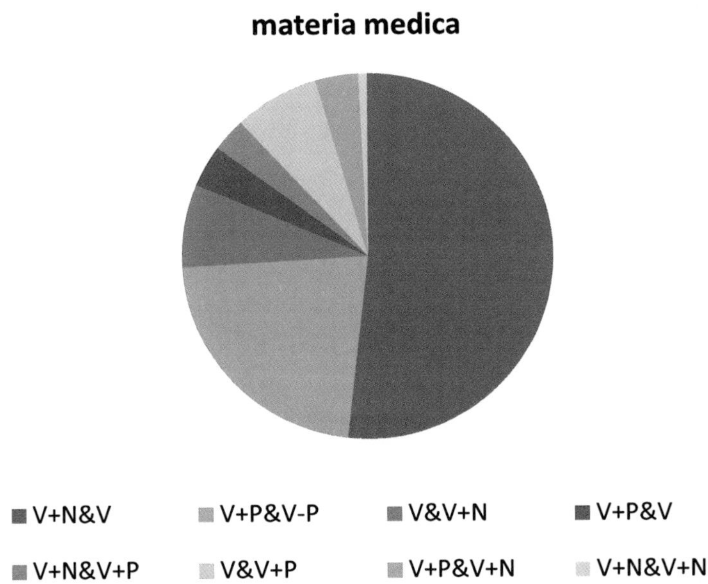

Figure 3a: Distribution of frequencies per tradition: *Materia medica*

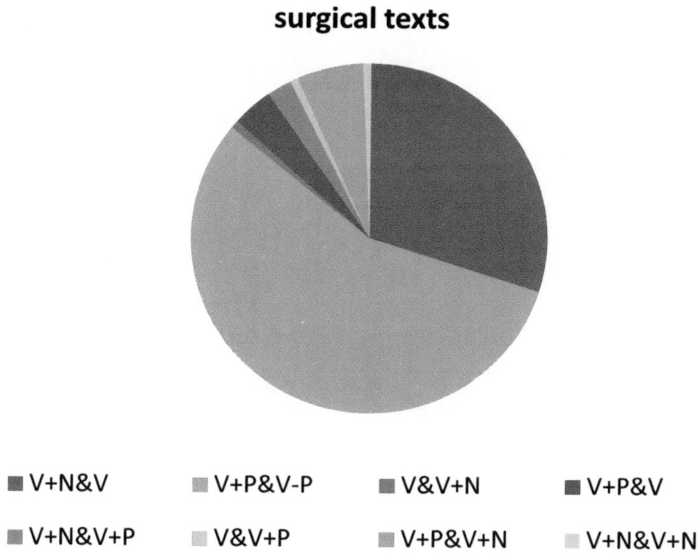

Figure 3b: Distribution of frequencies per tradition: *Surgical texts*

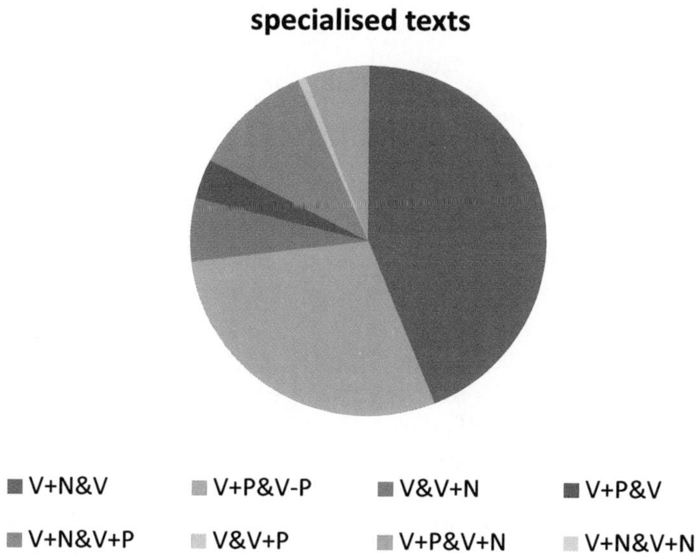

Figure 3c: Distribution of frequencies per tradition: *Specialised texts*

As already pointed out, all traditions present a high frequency of noun-pronoun and pronoun-pronoun object correlation. Remedies and *materia medica*, and surgical texts, which represent the two poles in the taxonomy of medical texts, show almost the same frequencies for these two structures but with opposite results. Remedies and specialised texts have similar frequencies of noun-pronoun objects. Specialised texts are similar to remedies, although noun-object correlations present fewer occurrences. Other structures with a significant frequency are zero anaphors with a pronoun in the first clause and zero anaphor with a noun in the first clause, especially in remedies and specialised texts.

5. Discussion

I will now concentrate on the description of each of the syntactic patterns mentioned above and the motivation for the selection of one or another. These eight structures are grouped into five categories: (a) zero anaphora, (b) zero cataphora, (c) proform anaphora, (d) proform cataphora, and (e) noun plus noun.

As I have already remarked in section 2, there are different approaches for the study of the presence or absence of complements, namely syntactic, pragmatic, and discursive. Culy (1996) includes for the first time the semantic variable for the interpretation of null complementation in recipes. However, this semantic component alone seems to be of little significance for data from the Middle English period. Givón (1983) proposes a discursive approach which consists in measuring the distance held between the anaphoric device and its referent, i.e. the lookback feature. This feature assigns values from 1 to 20 depending on how far the proform is from its referent.

Source of texts is also one important factor to take into account, for it gives an idea regarding personal or stylistic motivations behind the use or omission of nouns and referential devices. Finally, Culy (1996) points out that the grammatical factor is significant in the sense that *tense* may be held responsible for the presence of the direct object. A tensed verb normally carries an explicit object whereas the imperative seems to be less restricted. In my discussion, and in the conclusion, I would make manifest my intention to reconcile some of the existing views towards null complements within a cognitive approach.

5.1. Anaphora

Zero anaphora is represented unevenly in our corpus. Examples are the following:

1. Take peletre and sethe ____ in wynd (Medical works).
2. Make a plaster and leye ____ vp on the sore (Medical works).
3. Take jws of sentorye a good del and sethe ____ in good old ale (Stockholm).
4. Take it from the fyre and pwt ____ yn to a lytyl camfory (Maystyr peers).
5. Thanne put it in a clene glas vessell; and wete ____ wel that it hath the same vertu (Medical works).
6. Leye it in eysell and smere ____ with thyn hede al aboute (Medical works).

These examples are the proof of a potential tendency in Middle English to reduce the amount of redundancy in short simple sentences. In all cases, the type of verbs are the imperative mood and their semantics is defined in terms of the stage of the recipe progress in which the verbs appear. This model is proposed by Culy (1996: 94) for the specific case of the recipe and categorises verbs into (a) finished (subject of the recipe), (b) almost done (no further ingredients to be added), (c) working on (stages from beginning to almost done), and (d) other (this includes ingredients, pots, etc). The instances above present cases of working on verbs, as in *pwt*, almost done represented here by the verb *sethe*, and finished, that includes the forms *leye*, *wete*, and *smere*. The probability is 1:1:3, the most frequent being the finished one. Admittedly, this lacks representativeness because this is a large corpus and only few data are randomly taken into play, but the comparison of these figures partly confirms Culy's findings regarding the semantic factor. She argues, "the two groupings differed with respect to the properties of Finished and Almost Done, though in most cases Almost Done favored zeros/anaphors more than Finished" (1996: 98-99). Recall, however, that her historical survey has been performed on sixty recipes, twenty of which belong to the Middle English period. Her survey contrasts Görlach's (1992); he claims that "zero is quite uncommon... and [its use is] restricted to an object placed between two verbs" (1992: 47).

The following chart gives the results for the two groups of zero anaphora. It can be seen that specialised texts and remedies share a similar frequency in the case of noun-zero anaphora, whereas, in the case of pronoun-zero anaphora, the figure for Remedies exceeds the figure for specialised texts in something less than .65 points. The figures for surgical texts are not significant in comparison with the other two textual traditions.

Relative frequency of zero anaphora in the corpus

	V+N and V	V+P and V	V and V+P
ALL (MEANS)	0,47	0,36	0,17
REMEDIES AND MATERIA MEDICA	0,50	0,77	0,40
SURGICAL TEXTS	0,06	0,02	0,04
SPECIALISED TEXTS	0,58	0,12	0,04

Pronoun anaphora consists in the presence of a proform functioning as object of either one or the two verbal phrases, thus: V+N and V, V+P and V, and V and V+P. The distribution is, as follows:

Relative frequency of pronoun anaphora in the corpus

	V+P and V+P	V+P and V	V and V+P
ALL (MEANS)	1,80	0,36	0,17
REMEDIES AND MATERIA MEDICA	3,04	0,77	0,40
SURGICAL TEXTS	0,71	0,02	0,04
SPECIALISED TEXTS	1,13	0,12	0,04

The relative frequencies of columns 2 and 3 have been already discussed for the case of zero anaphors since these structures present both zero and pronoun anaphoric devices. The first column represents the instances where complementation in both verbal phrases is occupied by a proform. This structure is more common in the three traditions with a higher frequency in the case of remedies and *materia medica*. Examples are the following:

1. Then medyll them wyth ye grece and grynd them well (Medical works).
2. Set it ouer the fyr and thanne steme hit a litil (Medical works).
3. Tempere it with venegre and drinke it (Macer).

5.2. Cataphora

Zero cataphora is rarely studied or even mentioned in studies on null objects in English, probably because this is extremely infrequent, as can be seen in the relative frequency table, below. With respect to cataphora, Ohlander (1943-1944: 112) has argued that "examples of the type *He took out and lit a cigarette* are rare in the oldest period. Not until ME do they become more numerous".

Relative frequency of cataphora in the corpus

	V & V+N
ALL (MEANS)	0,33
REMEDIES AND MATERIA MEDICA	0,30
SURGICAL TEXTS	0,23
SPECIALISED TEXTS	0,40

The cataphoric device is unevenly distributed in our corpus, with a lower frequency in surgical texts, almost .17 points with respect to specialised texts. Remedies and *materia medica* is halfway between the two. Examples of cataphora are:

1. The fyre wastyth ____ and stroyeth the watyr (Wyse book).
2. It mollysyeth ____ and dyssolwyth hard Empostymes (Wyse book).
3. Þis, forsoþe, swageþ ____ and softeneþ þe brynnyng (Ardene fistula).
4. The fyrst cote he callyth "saluatricem," that is a sauyour, for it sauyth ____ and kepyth the humors (Ars oculorum).

The identification of this structure as having zero cataphora is controversial since, for many, the *and*-construction makes the verbal phrases as a single unit, and so zero cataphora would not be just possible. However, this claim is certainly weak as, from both a syntactic and a discursive perspective, the reader must go forward in the text to identify the object of transitive verbs, such as *wastyth* ('waste'), *mollysyeth* ('mollify'), *swageþ* ('mitigate'), and *sauyth* ('save'). A primary difference with anaphora is the use of tensed verbs, all of them inflecting for the present. Another salient characteristic is that zero cataphoric devices tend to appear in descriptive passages rather than instructive, but they match anaphora in that both are likely to appear with verbs meaning sort of a process outcome, whatever contents this "process outcome" refers to.

5.3. Noun plus noun structure

This pattern is relevant for my study in the sense that it prevents the presence of anaphora or cataphora. It implies the repetition of the same noun in a coordinated sequence favouring redundancy. This redundancy seeks in turn to avoid ambiguity and vagueness and to enhance clarity and objectivity, as in the examples given below.

1. as terenbine to moiste bodies; & to drie bodies (Chirurgia magna 1).
2. opene the place & exclude þe quytture, & clense al þe place (Wyse book).

Arguably, this structure may be compared in function with the type *take and peel the apple*, since the object is clearly contextually defined. In spite of this, I think that, behind the repetition of the same nouns, there must be an authorial intention to be precise and, consequently, reduce misunderstandings in the interpretation of the message. This is something the structure exemplified in *take and peel the apple* cannot possibly aim at. Occurrence in the corpus is restricted to the very few examples in *Chirurgia magna 1*, a surgical text, and *The wyse book of maystyr peers of salerne*, a remedy book.

6. Further thoughts on null complements

As already said, zero anaphora has been studied from different perspectives, from grammar to register. This absence of pronouns is not always predictable, not even by the factor of position within the text, as Culy (1996) tried to make manifest from the semantic component. Register is also used to explain this absence, but I am not fully convinced that register (in the sense given in Halliday and Matthiessen (2004)) is enough to explain zero complementation; in my opinion, that is more a description of facts than a motivation. A cognitive approach could give us a more suitable answer to the omission of zero anaphors, as earlier pointed out by Groefsema (1995) and Stainton (1997).

Could cognition cope with zero anaphor and cataphor? I firmly believe so. It is not so much a question of style, though it might seem plausible at first sight. In our corpus, zero cataphor occurs frequently between two verbal phrases strongly connected in meaning, as in the following:

1. mollysyeth and dyssolwtyth hard Empostymes (Wyse book of maystyr peers of salerne).
2. take and seþe verueyn (Medical works).
3. destruye and swage þe ache (De viribus herbarum).
4. stampe and medle þe juys of fenel (De viribus herbarum).

5. purge and clense þe face of freknesse (De viribus herbarum).
6. reteigneth & withholdith the propirte of hem (De spermate).
7. growith and thrivith the strength of the humour (De spermate).

These examples make evident either that the second verb refers to an action that depends on the realisation of the first action, as in *destruye and swage þe ache*, or that the second verb describes the same action as the first action, as the first verb, and its use is that of strengthening its meaning, as in *reteigneth & withholdith*. These conceptual repetitions have evident cognitive effects on the reader, as the writer seeks to reduce the level of cognitive efforts in the processing of a particular linguistic string including meanings other than explicitly stated (see Relevance Theory principles in Sperber and Wilson 1995). Style would be a consequence of this, rather than an end.

The motivations behind the use of zero anaphor are very difficult to ascertain, since the contexts of occurrence are varied. There is also a tendency for preposition stranding because of the object omission, as in *take peletre and sethe ____ in wyn* (*medical works*). The ellipsis of the object in contexts like this seems to be register-constrained, as detransitivisation is identified as typical of instructional texts. In our corpus, I have detected the use of covert pronominals in descriptive contexts such as *it tornyth that and fastenyth in the propirte of humours*. The reason for this omission is far from clear and, certainly, it does not seem to be driven by register or even genre. It could be that the first action with its object has a mental representation pervading during the realisation of the second action. In other words, the existence of this second activity relies on the existence of the first. This existential linkage is cognitively manifested in the elision of the pronominal object.

Let us examine the following two examples to see what the main differences between them are as for the presence/absence of complementation of the verb *ete(n)*:

1. Suffren hym to eten ____, y seyde þowȝ þat he wolde eten ____ he scholde noȝt (Chirurgia magna 1).

> 2. Who so wil kepe his teth fro akynge vse not to ete freshe wat~ fishe for þei gend~ lyghtly corrupcion abowte þe teth. Also bewar to ete cold mete after ryght hote or hoot aft~ cold he vse not moch raw onyons and lekys and kepe the wele fro brakynge if yow may for all þies done moche harme to þe teth and gendren lyghtly corrupcion of which comyth akynge (Leechbook 1).

The first instance shows the absence of complementation for the verb in question, whereas all the cases of *ete(n)* in the second instance carry a phonologically realised object: *to ete freshe wat~ fishe*, *to ete cold mete*. The motivations for the absence in the first case are varied, although none of them are mutually exclusive:

1. The omission or non-realisation of the object shows the interest of the speaker to focus on the process, rather than on the patient. Thus, absence is fully intentional to achieve a communicative goal, and complementation of the verb *ete(n)* refers to any type of food.
2. The omission is triggered by collocation. Here, collocation should be understood as a cognitive strategy by means of which the reader, given the appropriate contextual conditions, knows the complement without any further spelled-out instructions: eat meat, eat fish, eat (raw) vegetables.
3. The omission is contextually defined. Information on what is eaten must be retrieved from what is said before and after in the running text. In our last examples, context is useless to identify the reference of zero anaphora since we cannot find indications as for the food the patient is instructed to have. Therefore, again, this absence of the object may refer to any type of food. By context, I also refer to both the linguistic context in which zero complementation is embedded, and the physical/environmental settings in which an utterance takes place.
4. The omission is culturally defined. This is rather a complicated turn because the absence of the complement is due to what speakers and readers may be entitled to know just because they belong to (a) the same culture, and (b) the same group of

experts. In the first case, the complementation for *ete(n)* may refer to a particular meal in the day, probably dinner. The second case is one in which participants in the exchange share mutual knowledge about a same reality. This implies that reference is based on what is known to be adequate by the participants, and omission is made possible for this "mutually shared knowledge". Arguably, this motivation could be said to belong to context, and that is true, but I have preferred to distinguish between the context given by the text itself and the context given by the cultural settings. In this group, I also consider the concept of genre, as genre is a social and cultural construct that may help us understand the missing chunk (as in Culy 1996: 114).

All of these are cognitive in essence but especially motivations 1, 2 and 4. A random search in 59 MEMT texts shows that the *focus on the process* strategy serves the purpose of directing the attention of the reader on the action of eating rather than on the product eaten since this could be any. Observe the following instance: *No man shulde ete til he knew that his stomak wer voyde of that he had etyn beforn* (*Seven herbs, seven planets*). The complementation of the verb *ete* here seems to be elided in favour of clarity to avoid redundancy and vagueness in speech. If a list of product is given, it takes the risk of not being exhaustive, and so the verb *eat* not being complemented would epitomise the whole of edible things.

Regarding *collocation*, this search of the MEMT corpus offered different results for the verb *to eat*. Apart from some cases of zero anaphor, the verb *ete* is normally complemented with nouns, adverbs and pronouns. The most frequent forms are (a) nouns: *mete, metes,* and *metys* ('food'), (b) adverbs: *wel(e)* and *lytull/lytell*, and (c) pronouns: *it, hem*. Other frequent nouns are: *erbys, eggis, flesche, potage, hennys*. This strategy in combination with context and culture favours the absence of zero complementation in some other parts of discourse. As said above, the cases of null complements refer to general type of food, and hence specificity is avoided.

We have extensively seen how context plays its role in the disambiguation of zero complementation. My last bet is on culture. The following spoken examples taken from various sources will illustrate this point:

> 1. *Yes. We used to get up seven o'clock in the morning you know. Mm. Would you as All the servants would you eat together yourselves? Oh yes. Yes. Where would you eat, in the kitchen? Yes in the kitchen yes. Mm.* (British National Corpus, K6U).
>
> 2. *Creo que has exagerado demasiado esta ruptura porque yo voy todos los domingos a comer a casa – A comer, ¿eh?, a comer ¡Claro!, va a comer Sí, de vez en cuando hay que volver a la comida esa - de mamá y esos platos maravillosos* (Corpus del español, 19-0).

These two examples represent the seemingly intransitive behaviour of the verb *to eat/comer* in English and Spanish, respectively. Scholars have looked at this absence of complementation of this verb in two ways. A first one is to consider the verb *to eat* as a transitive verb with two valencies Agent and Patient. Syntactically, the feature Patient is elided whenever context makes it easy to retrieve the omitted information. A second view is to consider the verb *to eat* as an example of intransitive verb. Iten et al. (2006) introduce the notation [±add a complement] to the semantic frame of verbs following Fillmore's claim (1986: 97) that a lexical item may condition the absence or presence of complementation according to a [±omit a complement] feature (Iten et al. 2006: 6). Junker et al. (2004) specifically concentrate on the verb *to eat* as intransitive and they claim what follows:

> There is now the intransitive 'eat', which is what we see when the verb appears alone, and in addition the transitive verb, which is what appears with an overt complement. Continues the objection, this is unacceptable on two fronts. First, specifically with respect to introducing the intransitive 'eat', whenever someone eat; they eat something; but, goes the worry, treating 'eat' as intransitive we will miss this fact. Second, with respect to both entries, we are positing an ambiguity without necessity here. Our reply to the first worry is that this fact, i.e., about eating requiring that something be eaten, is real enough, but is not reflected in the semantic frame of the intransitive verb. The thing-eaten is not a linguistically inherent participant, the verb being intransitive,

> but it is a "metaphysically" inherent one – and that's enough to capture the fact in question. (To see the reasonableness of this suggestion, compare: whenever someone eats, they eat somewhere; that, however, is patently a poor reason for thinking that 'eat' is linguistically marked as expressing a three-place relation between an agent, a thing-eaten, and a location.) Our reply to the second worry is that there is no ambiguity posited without necessity, because we do not believe that there is a transitive verb 'eat'! Instead the sole semantic frame for 'eat' is (29):
> 29. a. Verb: 'eat'
> b. Semantic Frame: [Participant = Agent]
> This, in turn, raises an obvious question: How, then, can 'eat' take a complement, if it's always intransitive? 'Die' and 'fall', genuine intransitives, cannot. The answer is that 'eat' is an intransitive verb which is marked for optional addition of an argument.

It seems perfectly possible that the verb *eat/comer* may be intransitive with the option to be transitive by the semantic feature [±add a complement] in contrast to Fillmore who formulates the feature [±omit a complement] for verbs like this one, as said before. The first option seeks to resolve the problem of understanding the utterances given for speakers K6U and 19-0. If *to eat* and *comer* are intransitive verbs then no wonder these verbs are not complemented in these sentences. Junker et al. argue that what is eaten is not "a linguistically inherent participant" but a ""metaphysically" inherent one". In my opinion neither the verb *to eat* nor *comer* is intransitive. There is nothing metaphysical in the thing-eaten as a participant; it is mostly physical and bound to real life to the point that the action described by the verb *to eat* (or *comer*) requires, at least mentally, the inputting of some stuff to start the action of eating. This does not mean that linguistically we have to overtly express the thing-eaten, but that does not make the verb intransitive as far as we can retrieve from context or culture the object, including a generic one.

Junker et al. also say that the fact that Location does not appear in sentences containing the verb *eat* proves that the direct object is not omitted but optionally not taken. They say that, if someone eats something, this action of eating must be performed at one specific place. However, I cannot see the link in their claim. That circumstantial complements are not given in a sentence or in the semantic frame of the verb does not imply a lack of the direct object in that semantic frame. It is true, though, that we can add as much information about a particular event as we want to, but talking about transitivity is a different task. We could obviate the location, and even the time of process, in our mental representation of the action of eating; however, we cannot ignore the fact that the person eats something and that is linguistically inherent to the verb *to eat* and *comer*. Another thing is that phonologically the complement is not realised.

Another reason that comes to my mind not to see clearly the suitability of the verbs *eat/comer* as intransitive with complements optionally being added is that, by definition, an intransitive verb cannot present phonologically or mentally realised objects, unless we are talking about two different verbs having the same phonological form and the same epistemic meaning. However, it is possible for transitive verbs to realise or leave their complements phonologically non-realised. As I said before, transitivity is something else than the purely linguistic realisation of the complement. For this reason, I still consider Fillmore's claim (1986: 97) of the omission of the complement more suitable to describe the English and Spanish utterances given above and reproduced again below for clarity. We can explain this omission from a pragmatic point of view following the motivations mentioned above, namely: (a) focus on process, (b) collocation, (c) context, and (d) culture.

1. *Yes. We used to get up seven o'clock in the morning you know. Mm. Would you as All the servants would you eat together yourselves? Oh yes. Yes. Where would you eat, in the kitchen? Yes in the kitchen yes. Mm.* (British National Corpus, K6U).
2. *Creo que has exagerado demasiado esta ruptura porque yo voy todos los domingos a comer a casa - A comer, ¿eh?, a comer ¡Claro!, va a comer Sí, de vez en cuando hay que volver a la comida esa - de mamá y esos platos maravillosos* (Corpus del español, 19-0).

The complementation in the English example is easily retrieved from both the linguistic context and the context of culture. In the text, we are said the exact moment of the day in which the action takes place, i.e. seven o'clock. That means that what they eat is their breakfast during weekdays. The Spanish case is different in the sense that *comer* in *yo voy todos los días a comer a casa* means a very specific moment of the day. Culturally, Spaniards start their main meal between two and three in the afternoon and, apart from other expressions, it is very common to say 'voy a comer', '¿qué hay para comer?' or 'es la hora de comer' referring to have lunch. So, to utter *comer* in *yo voy todos los días a comer a casa* means nothing else than "eat your lunch". Lack of specificity as to what exactly has been eaten does not imply intransitivity. Consider, for instance, the utterance: *Cómete la comida* 'eat your meal'. *Comer* here is transitive, but still we do not have an idea as for the thing-eaten.

A search for the word *eat* in my corpus of study has revealed that this verb is always complemented with a direct object, except occasionally where zero anaphora takes place. In all those cases, reference has always been easily retrieved by reading back in the text. One such case is: *For why who þat etith withoute desire or appetite, his mete shal fynde the kyndly hete of his stomak cold and if he ete ____ with desire he xal fynde the kyndly hete of his stomak yeue up a smell like to encens (de ingenio sanitatis*). In this example, the word "ete" appears with a phonologically non-realised complement. However, this complement is contextually defined, and refers back to "his mete". There are some cases in which the zero anaphora is the consequence of what Roberge (2003: 6) calls *binding from the left periphery*: *strong ale, and benes, and peesen, ete þou not* (*Gilbertus anglicus compendium*). The reference precedes the verb due to an alternation of the syntactic structure in which the object is expected to follow the verb.

7. Conclusion

In this chapter, I have discussed the motivations lying behind the use of zero anaphora in Middle English medical texts. Earlier approaches have considered zero anaphora as the result of syntactic, pragmatic and discursive strategies in the text, even from a cognitive perspective. In a same line of thought, zero complementation has involved scholars in reworking the taxonomy of verbs. For this reason, verbs that have been traditionally transitive have now been reanalysed as intransitive according to their semantic frames. Junker et al (2004) and Iten et al. (2006) propose this way of analysing the verb "to eat", and so this verb comes to be transitive due to an optional feature. This contrasts with the analysis put forward by Fillmore (1986) in which complements can be omitted in transitive verbs. With all reservations, and granting the intransitivity of the verb "to eat", I have pointed out the possibility of having two types of this verb. Only with two forms can I understand that a naturally intransitive verb could sometimes appear as a transitive one.

I have demonstrated that, in the particular case of the verb "to eat", this is essentially transitive in Middle English. There are, however, few occasions in which this verb appears without an overt complementation but even then, the referents are clearly identifiable contextually. Another conclusion drawn from the present study is that this lack of isomorphism between semantics and phonology has four possible cognitive motivations; namely, *focus on process*, *collocation*, *context*, and *culture*. I have shown that all the cases in my corpus fall within at least one of these four cases.

A final contribution of this study is that frequencies of zero anaphora and cataphora devices are larger in the case of specialised texts. Pronominal anaphora and cataphora show a different distribution; the former presents a larger frequency in specialised texts and remedies, in this order, whereas the latter shows an even distribution in these two traditions. None of these devices seems to be characteristic of surgical texts. Overall, there appears to be room for more research to see the degree in which the tradition of writing, register and genre impose some constrains to the licensing of null complements. Recall that these three features would belong to what I labelled as *culturally defined anaphora*.

References

Alonso-Almeida, Francisco 2008. The Pragmatics of and Conjunctives in Middle English Medical Recipes. A Relevance Theory Description. *Journal of Historical Pragmatics*. 9/2, 171-199.

Bender, Emily 1999. *Constituting Context: Null Objects in English Recipes Revisited.* U. Penn. Working Papers in Linguistics. 6/1.

British National Corpus (UK 1980-1993). Available at <http://corpus.byu.edu/bnc/>. Page created by Mark Davies.

Carroll, Ruth 1999. The Middle English Recipe as a Text-type. *Neuphilologische Mitteilungen*. 100/1.

Corpus del español. Available at <http://www.corpusdelespanol.org/>. Page created by Mark Davies.

Culy, Christopher 1996. Null Objects in English Recipes. *Language Variation and Change*. 8.

Fillmore, Charles 1986. Pragmatically Controlled Zero Anaphora. *Bulletin of the Linguistic Society*. 12.

García-Velasco, Daniel / Portero-Muñoz, Carmen 2002. *Understood Objects in Functional Grammar.* Working Papers in Functional Grammar. 76.

Givón, Talmy 1983. *Topic Continuity in Discourse: A Quantitative Cross-language Study*. Amsterdam: John Benjamins.

Görlach, Manfred 1992. Text-types and Language History: The Cookery Recipe. In Startvik, J./ Wekker, H. (eds) *History of Englishes. New Methods and Interpretations in Historical Linguistics.* Berlin and New York: Mouton de Gruyter.

Groefsema, Marjolein 1995. Understood Arguments: A Semantic/Pragmatic Approach. *Lingua*. 96.

Halliday, Michael / Matthiessen, Christian 2004. *An Introduction to Functional Grammar.* London: Hodder Arnold.

Hickey, Raymond 2003. *Corpus Presenter*. Amsterdam and New York: John Benjamins.

Iten, Corinne / Junker, Marie-Odile / Pyke, Aryn / Stainton, Robert / Wearing, Catherine 2004. Null Complements: Licensed by Syntax or by Semantics-Pragmatics? *Proceedings of the 2004 Annual Conference of the Canadian Linguistic Association.*

Junker, Marie-Odile / Stainton, Robert / Wearing, Catherine 2006. *The Semantics and Syntax of Null Complements*. Carleton University Cognitive Report.

Ohlander, Urban 1944-1945. Omission of the Object in English. *Studia Neophilologica.*

Roberge, Yves 2003. Transitivity, grammaticalization and change. Linguistic Variation in Bilingual Communication, Acquisition, and Diachronic Change. Available at <http://www.chass. utoronto.ca/~roberge/pdf/Hamburg2003.pdf>.

Stainton, Robert J. 1997. Utterance Meaning and Syntactic Ellipsis. *Pragmatics and Cognition.* 5/1.

Sperber, Dan / Wilson, Deirdre 1995. *Relevance: Communication and Cognition.* 2nd edition. Oxford: Blackwell.

Taavitsainen, Irma 2001. Middle English Recipes. Genre Characteristics, Text Type Features and Underlying Traditions of Writing. *Journal of Historical Pragmatics.* 2/1.

——— 2006. Thought-styles and Metadiscourse in Six Hundred Years of Medical Writing. In Gotti, Maurizio / Salager-Meyer, Francoise (eds) *Medical Writing.* Bern: Peter Lang.

——— / Pahta, Päivi / Mäkinen, Martti 2005. *Middle English Medical Texts.* Amsterdam: John Benjamins.

Voigts, Linda Ersham 1982. Editing Middle English Medical Texts: Needs and Issues. In Levere, Trevor (ed.) Editing Texts in the History of Science and Medicine. New York: Garland.

——— 1984. Medical Prose. In Edwards, A. (ed.) *Middle English Medical Prose: A Critical Guide to Major Authors and Genres.* New Jersey: Rutgers University Press.

GRAHAM D. CAIE

The Hunterian Collection at the University of Glasgow

Glasgow University Library has one of the largest and finest collections of medieval manuscripts and early printed books in the world. The Special Collections Department has c. 300,000 manuscript items and over 1000 incunabula; the Hunter Collection alone has 650 manuscripts, over 250 of which are from the medieval period and many are unique. Facsimiles of a selection of these manuscripts can be seen at <http://special.lib.gla.ac.uk/exhibns/treasures/index.html> and <http://special.lib.gla.ac.uk/exhibns/chaucer/index.html>.

Work has begun with the digitisation of complete medieval vernacular manuscripts in the Hunter Collection, as can be seen at the website: <http://www.memss.arts.gla.ac.uk/>. The aim of the Digitisation of Middle English Manuscripts Project is to prepare digitised facsimiles of all Middle English manuscripts in the collection, accompanied by diplomatic editions of the texts and notes and where possible digitised images of early printed editions of these texts. *The Middle English Romaunt of the Rose* (Hunter 409), written in c.1440, and John Gower's *Confessio Amantis* (Hunter 7), c.1425, have been completed; these will be followed by Nicholas Love's *The Myrrour of the Blessid Lif of Ihesu-Crist* (Hunter 77), c. 1475, John Lydgate's *Fall of Princes*(Hunter5), c. 1431, and his *Life of Our Lady* (Hunter 232), Chaucer's *ABC* poem (Hunter 239), c. 1450, *The Chronicles of England* (the *Brut* is in six manuscripts in the Hunterian collection– numbers 61, 74, 83, 228, 230 and 443); two copies of John Wycliffe's *New Testament* (Hunter 189 and 176), and finally the 1479 Spireleng copy of Chaucer's *The Canterbury Tales.*

The Collection has the only surviving copy of the Middle English *Romaunt of the Rose*, a large part of which is considered to be Chaucer's own translation. Dating from the mid-fifteenth century, this codex was subsequently used by William Thynne as the copy text for his edition of Chaucer's works printed at London in 1532. There is also a copy of this edition in Special Collections and so it was digitised along with the manuscript and is displayed on the same website.[1]

1. William Hunter

First, who was Hunter and why did the University of Glasgow receive his collection which is one of the largest holdings of historical medical manuscripts in the world? Dr William Hunter (1718-83), doctor, Physician Extraordinary to Queen Charlotte was also a collector of anatomical and natural history specimens, as well as books and manuscripts. About one third of Hunter's books, as one would expect, are about medicine. The reason why his collections came to the University of Glasgow is because he started his studies there and bequeathed all his collections to his alma mater on his death. In 1807 they were housed in the old buildings of the University, founded in 1451, and so in 2007 it celebrated the bicentenary of the opening of the Hunterian Museum. This was the first public museum in Scotland and only the third in the UK after the Ashmolean and the British Museum.

Hunter was born at Long Calderwood Farm near Glasgow in 1718 and matriculated in divinity at Glasgow University at the age of 13; he later studied medicine at Edinburgh. At the old age of 22 he moved to London to join many other Scottish doctors there. He lodged with Dr James Douglas, who was not only an important royal physician, but also a great scholar and collector of books; indeed much of the Hunter collection comes from Douglas. In addition to this role model for William Hunter, was another famous London doctor who was also a collector, namely Hans Sloane (1660- 1753), an Ulster

1 See the digitised version at <http://www.memss.arts.gla.ac.uk/html500a/057v500a.htm>.

Scot, whose collection, combined with the royal library of George II, provided much of the material for the British Museum in Bloomsbury, London.

Hunter was elected a Fellow of the Royal Society in 1767 and Professor of Anatomy at the Royal Academy in 1768. He was a leading teacher of anatomy and created a lecture theatre in Leicester Square, London, to which Sir Joshua Reynolds and other artists would also come to study anatomy. From here he assembled a formidable collection of anatomical species which now forms part of the Zoological Museum in the University of Glasgow.

Hunter also collected works of art as well as coins, minerals, shells, anatomical and natural history specimens and all of these after his death in 1783 were bequeathed to Glasgow University, and are now housed in the Hunterian Museum; his art collection is in the University's Hunterian Art Gallery and his 650 manuscript codices are in the Hunterian Collection in Special Collections. In addition there are over 10,000 printed volumes in the collection, over 500 printed in the 15th century, ten of which were printed by William Caxton at Westminster.

Hunter's holdings of vernacular literature was just as impressive: Chaucer, Shakespeare, Spenser, Cervantes, Dante, Boccaccio, Rabelais are in the collection, as well as the writings of the great 16th- and 17th-century physicists, Copernicus, Kepler, Tycho Brahe, Newton and Hooke. Luckily we have Hunter's own catalogue of his library, begun four years before his death in 1783, and this contains information about how he acquired many of the works. Some were donated by friends and acquaintances; in 1774, for example, Samuel Johnson asked Hunter to present to the King a copy of his recently published *Journey to the Western Isles*, as well as giving one to Hunter himself. Adam Smith presented a copy of the second edition of his *Wealth of Nations* and Sir Joshua Reynolds gave him his *Discourses to the students of the Royal Academy*. But the bulk of Hunter's contemporary books were purchased directly from booksellers and at auctions. It is interesting to see that at this time early printed books, in particular incunabula, were more expensive than manuscripts. For example the magnificent and ornately illustrated and illuminated 12th-century Hunterian Psalter cost 50 livres and 1 sou, a fraction of the sum paid for some 15th century printed books.

2. Medical Manuscripts

As one might expect the majority of his book and manuscript collection comprises works of a scientific nature, largely medical. Hunter possessed two illustrated copies of the celebrated surgical treatise of the 14th-century surgeon, John Arderne (Hunter 112 and 251), who has sometimes been called the father of surgery. Ardenne was from Newark-on-Trent and Nottingham (1307?-1392) and learned much of his surgical craft during the Hundred Years' War, when he created an ointment to heal arrow wounds. One of his works was on the treatment of fistula in the anus, or abscess in the rectum, to which knights on horseback were prone. His *Speculum phlebotomiae* or *Mirror of Phlebotomy and Practice of Surgery* was frequently copied and includes the best examples of medical illustrations of the period. The frontispiece shows a physician with a ladle and medicine box sitting in a fenced herb garden. He also gives advice in this work on fees to be charged, namely that the rich should pay high fees and the poor very little. In one copy (Hunter 251) from the second quarter of the fifteenth century there is a wonderful Zodiac man, indicating to the surgeon which parts of the body were influenced by which astrological sign.[2] One can see that the feet are influenced by Pisces and the head by Aries. Many of these works are richly illustrated and plants and herbs are drawn to aid the physician make up his medicines.

For this current project, based at the University of Málaga, the following Hunterian manuscripts have been selected: 95, 328, 497, 503, 509 and 513. There is a growing interest in medieval scientific texts, but much of the material is inaccessible and unedited. This

2 See <http://special.lib.gla.ac.uk/exhibns/chaucer/medicine.html>.

collaborative project brings together the expertise of the researchers in Málaga, both in editing and in technological innovation, and the manuscript resources that are available at Glasgow University.

Hunter 95 (T.4.12) comprises a series of medical receipts by unknown authors and written in the fifteenth and sixteenth centuries. This manuscript is in both paper and vellum. The major items are: a Middle English translation in the fifteenth century entitled *Surgery* of the Latin work *Inventarium sive chirurgia magna*, which was completed in 1363 by Guy de Chauliac (ff. 34r-73v). Guy de Chauliac was an early fourteenth-century French surgeon who lived in Avignon and attended Pope Clement VI. On ff. 82r-156v is a fifteenth-century *Book of Operation*, which begins with a detailed index of the contents. On ff. 156v-85v is the fifteenth-century Middle English translation of Yuhanna Ibn Masawaih's *Antiodotary*. He was an Assyrian physician, born in 777 AD, who worked in Baghdad. He wrote medical treatises on a wide range of topics, including ophthalmology and fevers and medical aphorisms.

Hunter 328 (U.7.22) is another miscellany of medical texts of the fifteenth century in vellum. It contains a work by John Ardenne, who is described above (Hunter 112 and 251), which is a commentary on Aegidius of Corbeil's (also known as Gilles de Corbeil, 1165-1223) Latin treatise on urines, *Tractatus Metricus de Urinis* (ff. 1r-44v). This manuscript also contains two anonymous alphabetical lists of remedies on folios 45r-68v.

Hunter 497 (V.7.24) is a Middle English translation by John Lelamour (fl. 1373) of Macer's *De Virtvtibvs (sive Viribvs) Herbarvm*, entitled *Herbury*. It also has many sixteenth-century marginalia.

Hunter 509 contains the Middle English adaptation of Gilbertus Anglicus's *Compendium Medicinae,* an encyclopaedic work written c. 1240; the vernacular version in Hunter 509 is dated c. 1400 and the first printed text was in 1510. It was one of the most important medical works and the English adaptation was copied and used until the seventeenth century. Chaucer's Doctor of Phisik mentions "Gilbertyn" in his list of famous physicians.

Hunter 503 (V.8.6) has the Middle English translation of Benvenutus Grapheus's work *De Probatissima Arte Oculorum*, entitled *A Treatise on the Use of the Eyes.* Benvenutus Grapheus, sometimes called Grassus, was an Italian physician of the thirteenth century, probably of the famous medical school of Salerno. The Hunter manuscript is in a very fine textura semi-quadrata hand with coloured capitals and marginalia from the sixteenth century. L. M. Eldredge (1996) has edited this work, using Hunter 503 as base text.

References

Eldredge, L. M. 1996. *The wonderful art of the eye: a critical edition of the Middle English translation of his De Probatissima Arte Oculorum*. East Lancing: Michigan State University Press.

JAVIER CALLE-MARTÍN

Line-final Word Division in Late Middle English *Fachprosa*: G.U.L. MS Hunter 497 (V.7.24)[1]

1. Introduction

Even though word division at the end of a line is found in almost any piece of written material, its study in Middle English manuscripts is often cold-shouldered by scholars and editors on account of the apparently erratic practice of scribes. In fact, this is a topic tip-toed around in almost every palaeography publication, wherein its exclusion seems to be the rule rather than the exception (Johnston 1906; Osley 1980; de Hamel 1992; Derolez 2003; Roberts 2005).

The only references found in the literature leave the reader in a welter of confusion as these accounts are often contradictory. On the one hand, the traditional tenet considers that there is no convention, neither at a morphological nor at a phonological level, governing word division among Middle English scribes. The phenomenon is then considered as haphazard in the handwriting of the period where the only precept "seems to have been that not less than two completing letters could be carried over to the second line" (Hector 1958: 48; Petti 1977: 31; Denholm-Young 1964: 70). This traditional view, however, has been discredited in the last decades in light of the analysis of other mediaeval witnesses. In this vein, Lutz observes that the "scribal division of polysyllabic words reflects their phonological organization

1 The present research has been funded by the Autonomous Government of Andalusia (grant number P07-HUM02609) and by the Spanish Ministry of Science and Innovation (grant number FFI2008-02336). I am also grateful to the Caledonian Research Foundation - Royal Society of Edinburgh for generous financial support (2006 European Visiting Research Fellowship). These grants are hereby gratefully acknowledged.

into syllables [and] it can be therefore assumed that word division at the end of lines in Old English is based on the syllabification of that language" (Lutz 1986: 193; Burchfield 1994: 182). Hladký, in turn, offers the other side of the coin arguing that the main word division principle in Alfredian Old English is basically morphological, although he further recognises the increasing importance of phonetic division throughout the Middle English period (Hladký 1985a: 74-75; Hladký 1985b: 137).

Notwithstanding the validity of these recent approaches, they present some shortcomings which cast doubts upon the level of scribal standardization. Although examples proliferate in these publications, they are more often than not limited to listing the word-breaking patterns found, and the reader never gets to know the frequency and consistency of a particular type of word boundary, shunning the subsequent comparison with other prose texts, either synchronically or diachronically. Therefore, an empirical study of word division is still a *desideratum* in early English manuscript composition insofar as the statistical analysis of the phenomenon in a comprehensive corpus would shed light on the particular choices of the writer and, more importantly, it would eventually afford conclusive data about the level of expertise of some Middle English scribes in terms of word-breaking.

In view of this, the main objective of the present paper is the study of line-final word division in late Middle English in order to a) classify the type of splitting, whether morphemic or phonological; and b) offer a taxonomy of the main word-division rules from a statistical perspective. This approach will allow us to characterise the concept of syllable boundary in the vernacular together with its orthographic implications and, on the other hand, to validate or refute the traditional tenet about the afore-mentioned scribal inconsistency. Accordingly, the present paper has been organised into four different sections. Section 2 deals with the methodology and the theoretical framework; section 3 surveys word-division usage in the text under scrutiny in terms of the punctuation symbols used and the rules involved; finally, our conclusions are accounted for in section 4.

2. Methodology

The present research stems from a major project developed at Málaga University that, in collaboration with the Universities of Glasgow and Oviedo, pursues the digital edition of the mediaeval *Fachprosa* housed in the Hunterian Collection at Glasgow University Library. Our impulse is the preparation of on-line editions wherein manuscript images are accompanied by their corresponding diplomatic transcriptions. The project also tackles the compilation of an annotated corpus wherefrom the user may search for the occurrences of particular items, both word and lemma-based, context and accidence included. For this purpose, the annotation includes the lemma following the *e-MED* (Kurath et al. 1956-2001), accidence and meaning of the words in the corpus.

For the present study, the source of evidence consists of MS Hunter 497 (henceforth H497), containing a mid to late 15th-century English version of Macer Floridus' *De Viribus Herbarum*, ff. 1r-92r (Young and Aitken 1908: 409; Cross 2004: 33). This herbal is divided into 77 chapters wherein the author describes the medical qualities of herbs giving a comprehensive account of the diseases that may be cured by the medicinal receipts provided by these herbs (Keiser 1998: 3642).

The text is written by a sole hand in vellum using a legible 15th-century cursive *Secretaria* script, with a minute influence of the *Anglicana* — letters <s> and <w> in particular (Petti 1977: 14-15; Derolez 2003: 142-153; Roberts 2005: 211-213). The use of such a neat handwriting, together with the existence of conventional margins and frame ruling leads us to infer that the compositor was every now and then worried about the final layout of the text to provide a visually appealing product. H497 has therefore been chosen on an orthographic basis, as the author is meticulous to write until the verge of the frame, which motivates the existence of a wide variety of word division instances. The text under scrutiny stands out as an appropriate input for the analysis of line-final word division in late Middle English since it amounts to 35,950 running words and 954 word-division instances (one out of 37.5 words is affected by this kind of splitting).

Methodologically speaking, the digitization process was mandatory for subsequent transcription, which was diplomatic in all cases in order to preserve lineation as in the original. Irrespective of the scribal punctuation symbol, all word-breaks were provided with a hyphen to ease the automatic retrieval. *WordSmith Tools 5.0* (Scott 2008) was eventually used to list and arrange alphabetically the words containing the hyphen. Next, the instances were taken to an Excel spread-sheet to facilitate their statistical analysis and any subsequent arrangement according to the type of splitting involved.

The rationale used for our investigation derives from Hladký's study of word division in some historical texts (1985a; 1985b), wherein he approaches the topic by considering the ultimate force of splitting. Kladký recurs to the traditional word-formation principles as word-breaking devices, such as prefixation, suffixation and composition. On a phonological basis, he observes the frequent division after an open syllable, between two consonants, or between the pairs *-st* and *-ct*. Hladký's list has been supplemented with the inclusion of other word breaks to describe the scribal attitude towards word division in the witness under scrutiny.

3. Word division in H497

3.1. Punctuation

Scribal usage varies significantly across centuries and scriptoria. Whilst Old English scribes systematically deprecated the use of punctuation to mark off line-final word division (Burchfield 1994: 182), some heterogeneity is found in late Middle English insofar as the colon, the hyphen and the double hyphen coexisted for these purposes (Parkes 1992: 304; Petti 1977: 27; Calle-Martín and Miranda-García 2005: 32). The hyphen consists of a single oblique stroke, being the most frequent punctuation symbol in the 13th and 14th centuries. The double hyphen, in turn, spread from the latter part of the 15th century in English documents, consisting of two thin oblique strokes slightly curved upwards (Tannenbaum 1930: 146; Petti 1977, 27). In spite of the progressive standardization of punctuation symbols in these environments, they could often be left out where there was insufficient room at the edge of the folio, especially from the 13th century.

In the particular case of H497, the double hyphen is systematically adopted to signal line-final division. There are just a few instances where the punctuation symbol is missing, plausibly as a result of the writer's slip. The scribe of H497 makes the most of the writing surface and discards the use of line-fillers which become so frequent in other prose compositions of the period. This actually explains the high number of word-division instances in the text, if compared with other pieces. For illustrative purposes, G.U.L. MS Hunter 328 (henceforth H328), housing a vernacular version of Gilles de Corbeil's *Treatise on Urines*, amounts to 26,831 running words and 182 instances of word division (one out of 147.4 instances is then affected by a line-final splitting). More importantly, however, is the case of G.U.L. Hunter 3, a collection of 68 Elizabethan proclamations, wherein the use of line-fillers proliferate and word division is restricted to a couple of instances (Calle-Martín and Miranda-García 2008: 374).

3.2. Statistical analysis

As in the case of punctuation, word division is subjected to the personal idiosyncrasy of the scribe inasmuch as a particular word-division rule may become a recurrent practice in some texts whilst occasional or non-existent in others. This fact motivates the existence of drastic clashes when several prose pieces are compared, both quantitative and qualitatively. For comparison, Table 1 below reproduces the instances found in H497 and H328 both in absolute and normalized figures, the latter have been normalized to a text of 10,000 words to avoid text-length dependency (see Calle-Martín forthcoming for a detailed account of word-division usage in early English prose).

	H497		H328	
	Absolute	Normalized	Absolute	Normalized
Morphological division	68	18.91	25	9.31
Phonological division	872	242.55	136	50.68
Anomalous divisions	14	3.89	22	8.19
Total	954	265.35	183	68.18

Table 1. Number of instances.

Quantitatively speaking, we observe a major concern for line-final word division in H497 insofar as its occurrence becomes four times bigger than that in H328 (265.35 vs. 68.18, respectively). Even when the folios of both treatises comply with the requirements of the period as regards the margin standards and frame ruling, the author of H328 is many a time heedless of such boundaries, thus preferring to write beyond the actual frame, which reduces the number of word-division instances.

From a qualitative standpoint, on the other hand, phonological division is favoured in both witnesses as it occurs in 242.55 and 50.68 instances in H497 and H328, respectively. The author of H497 is particularly committed to use phonological division to the detriment of its morphological counterpart. Notably significant becomes the use of *humer-rus* to denote, according to the *e-MED*, 'one of the four fluids which forms the body', whose orthography presents the use of a non-etymological <r>, which is not tracked down in the other spellings of that word throughout the treatise. This consonant has been voluntarily added by the writer in order to avoid the morphemic division *humer-us*, thus fostering, at least visually, the use of a phonological boundary in *humer-rus*.

In light of this, we may conclude that the author of H497 shows a preconceived set of rules which are consistently spread throughout the entire composition wherein phonological division outnumbers morphological division, the latter depending on the room available for writing at the margin of the folio. H328, however, is less conventional, a fact leading to the use of cases like *d-rynke*, *w-ihte* or *bl-ack*.

According to this, the sections below classify the examples in H497 in terms of their breaking motivation, hence distinguishing two different parts. The first explores phonology as the pivoting force of word breaking. The second, in turn, concentrates on the morphological boundaries used by the writer —prefixation, suffixation and composition— and, in the case of the former, the prefixes and suffixes which are more frequently split.

3.3. Phonological division

A word in Middle English is taken to have as many syllables as vowel sounds. For instance, words like *sykness*, *natheless* or *phylosopher* are considered to have two, three and four syllables, respectively. More problematic seems to be the phonological implication of final *-e*, either etymological or non-etymological, in both native and non-native words. For chronological reasons, this vowel is considered to be mute in the 15th century in the case of native words like *muche*, *mouthe*, *throte* or *oute* with the exception of French borrowings, where the actual pronunciation of the vowel is sometimes retained as in *beute*, *myrre* or *quantyte*, among others. When necessary, we recur to the phonological dimension of final *-e* to determine whether it participates in the division of a given word.

As mentioned above, the author of H497 is meticulous as to write until the actual frame-ruling of the folio. However, unlike other mediaeval scribes whose practice may be deemed more erratic, he/she systematically avoids the stranding of a single letter in the second line, basically for aesthetic purposes. This leads to resort to both etymological and non-etymological *-e* in many contexts, even when it was not pronounced in 15th-century English. Examples actually proliferate throughout the text and a quick glance at the origin of some words, both native and non-native, reveals that the use of this vowel many a time depends on word division. Monosyllables perfectly illustrate the use of a non-etymological *-e* when they are broken at the end of a line, as in the cases of *yn-ne*, *ou-te*, *y-che*, *o-ne*, *te-the*, *wol-le*, etc. Irrespective of their number, monosyllabic adjectives are also prone to use final *-e* when they are broken at the end of a line, as in *why-te*, *leu-ke* or *ho-le*. The phenomenon also affects non-native words, as shown in *carewe-ye* vs. *carewey*, *lyly-e* vs. *lyly*, *mane-re* vs. *maner*, *betey-ne* vs. *beteyn*, etc.

This section surveys phonological word division in the data under scrutiny in order to determine, on a statistical basis, the resources available for a late Middle English scribe when breaking a word at the margin of the folio. There are 872 instances of phonological division in H497, which have been accordingly classified in terms of its orthographic environment, thus distinguishing whether a vowel or a consonant is involved. This procedure eventually helped us spot the different division boundaries, which basically respond to five major phonological rules: division after an open syllable, division between two consonants, division between two adjacent vowels, division between the pair *-st* and division between the pair *-ct*. The particulars of each are accordingly accounted for below.

3.3.1. Division after an open syllable (henceforth CV - CV rule)

It is the most widespread rule in H497 with a total of 457 instances, representing 51.58% of the occurrence in the corpus. This kind of division is more frequent in bisyllabic words (287 instances) than in three- and four-syllable words (130 and 39 instances, respectively). The preceding syllable may consist of a consonant plus a vowel as in *clary-fye*, *dely-uer*, *uene-mous*, or of a single vowel such as *a-boute*, *a-nother, o-dour*, *e-uer*, etc. The only exception to this rule is consonant <x>, which is systematically attached to the preceding vowel regardless of any other phonological consideration as in *lax-atyf*, *excyte*, etc. This orthographic convention often coincides with morphemic division, as in *relax-yth* and *wax-en*.

3.3.2. Division between two consonants (henceforth the C - C rule)

This rule consists in the separation of two neighbouring consonants when they appear at the end of a line. The scribe of H497 recurs to this rule in 372 instances (41.98%), of which 13 are monosyllabic and 293 are two-syllable words. Three- and four-syllable words, in turn, are obviously less frequent as there are 57 and 7 instances, respectively.

The examples have been classified in terms of their consonantal splitting to obtain the most frequent clusters. Table 2 below reproduces the different breaks in descending order. Occurrences are given in absolute figures.

Consonants	Total	Consonants	Total
n-k	34	n-s	8
r-t	32	r-g	7
n-d	28	d-d	6
r-b	23	c-k	5
n-t	21	l-k	5
s-s	19	r-r	5
l-l	18	d-l	4
n-ch	15	l-n	4
n-g	13	m-b	4
r-m	13	n-n	4
m-p	12	t-t	4
l-d	11	l-v	3
c-c	8	n-v	3
l-p	8	r-d	3
l-s	8	r-l	3

Table 2. C - C breaks in H497.

The most immediate conclusion that we can draw is that there is a major preference for word breaks containing a liquid consonant followed by a voiceless plosive, such as *n-k*, *r-t*, *n-t*, *m-p*, *l-p* and *l-k*, amounting to 112 instances, as in the following cases: *fran-kecense*, *par-te*, *perauen-ture*, *stam-pe*, *hel-pyth*, *myl-ke*, etc. Next, these are followed by clusters containing a liquid consonant plus a voiced plosive like *n-d*, *r-b*, *n-g*, *l-d*, *r-g*, *m-b* and *r-d*, amounting to 89 instances, as in *cor-yaun-dre*, *her-be*, *lun-ges*, *chyl-dryn*, *pur-gyth*, *wom-be*, *gar-deyn*, etc. The third most frequent splitting involves groups such as *s-s*, *l-l*, *c-c*, *d-d*, *r-r*, *n-n* and *t-t*, with 64 instances, as in *dys-sese*, *el-lybur*, *boc-ches*, *blad-der*, *myr-re*, *begyn-nyng*, *gobet-tes*, etc. Finally, clusters involving a liquid and a fricative consonant like *n-ch*, *l-s*, *n-s*, *l-v* and *n-v* are less frequent with 37 examples, like *staun-che*, *mul-sa*, *encen-se*, *sel-uer*, *syn-uey*, etc.

The only exception to this rule is the systematic preservation of any consonant followed by a liquid consonant providing that both of them belong to the same syllable, as in *de-gre*, *ga-dre* or *vyne-gre*, among others. This rule is elsewhere termed as the consonant + liquid rule (henceforth the CL rule) and its use is widely spread among Middle English scribes (Hladký 1985b: 136). In this same fashion, the writer of H497 consistently conforms to the CL rule insofar as no counterexample is found, adopting either the CV - CV rule (*togy-dre*), or the C - C rule (*gen-dryd*) instead.

Owing to the systematic shortening that affected closed syllables in early Middle English, the C - C rule is particularly found after stressed short vowels, with the exception of syllables affected by homorganic lengthening like *ol-de*, *gryn-de* or *wyl-de* and borrowings such as *her-be*, *com-fortyth* or *lar-ge*.

3.3.3. Division between two adjacent vowels (henceforth the V - V rule)

This rule consists in the separation of two neighbouring vowels at the end of a line on condition that they are pronounced separately. There are 23 examples in the data under scrutiny, representing 2.59% in the corpus. This rule is more widespread in the specialized terminology, usually of Latinate origin, as a clue for an appropriate reading, as in *siri-us*, *sci-etica*, *scla-yre*, *uari-ole*, *Recady-as*, *puly-oll*, *opi-um*, *uy-olet*, etc. Native words, however, are also occasionally divided according to the V - V rule, as in *sa-yth*, *sa-yn* or *bra-yde*. In this fashion, groups of three (or more) vowels also follow the V - V rule on condition that they are pronounced separately, as in *cory-aundre*, *noy-us* or *carewe-ye*, the latter with a non-etymological *-e*. This rule is only ignored in the event of morphological division, in which case the boundary is usually set after the second vowel, as in *destroy-eth*, *de-stroy-e*, etc.

By contrast, when the conjoined vowels represent a single sound, the boundary is consistently placed either before or after the vocalic cluster so as to ensure its pronunciation as a monophthong, and at the same time prevent the likely association of any of the vowels with the neighbouring syllables, as in *cou-loure*, *flou-re* and *regyoun*.

Notwithstanding the use of the V - V rule in H497, it is significant to mention that the writer may often opt for an alternative division, which consists in placing the boundary after the conjoined vowels, as in *entray-les* vs. *entra-yles*, or *kow-hers* vs. *co-whe*. This is an influence of the CV - CV rule, which was the dominant phonological rule in late Middle English manuscripts. The use of one rule or another actually depends on the particular needs of the writer in view of the space available at the margin.

3.3.4. Division between the pair -st (henceforth the ST rule)

This rule consists in the separation of the consonantal group *-st* to signal the syllabic boundaries of the word. There are 19 instances of the ST rule in the corpus, representing 2.14%, as in the following instances: *plas-ter*, *cos-tumably*, *des-troye*, *nos-trelles*, *fes-tre*, *has-tely*, *cas-te*, etc.

As shown, the writer is meticulous as to divide this cluster by considering that these consonants belong to different syllables. If a monosyllable, however, the ST rule is systematically disregarded considering that both consonants would then belong to the one syllable. The use of the C - C rule prevails over the ST rule in cases like *fyr-ste*, where the final vowel is non-etymological and without any phonological correlation.

3.3.5. Division between the pair -ct (henceforth the CT rule)

This rule presents the splitting of the consonantal group *-ct* to indicate that the consonants belong to different syllables. Unlike the ST rule, the use of the CT rule is found to be erratic among Middle English scribes as it is entirely ignored in some prose pieces (H328 for instance), whilst it becomes occasional and frequent in others (Hladký 1985a: 74). There is only one instance of this type of breaking in H497, i.e. *confec-te*.

3.4. Morphological division

Historically speaking, morphological division presents a decreasing level of importance in the sense that it was eventually superseded by phonological division after the arrival of the Normans (Hladký 1985a: 73-75). The obliteration of morphology as the guiding word-division principle runs parallel with the progressive recognition in Middle English that words should be split into syllables so that the reader may easily recognize the word just by reading the first part of it. When morphology participates in a given word-break, Powell considers that a conflict arises between the origin/meaning of a word and the natural grouping of sounds, "with the result that morphology, when easily recognizable, interferes with the phonetic division of the word" (Powell 1984: 456). This actually explains the historical preference for phonological division from the early Middle English period.

This section investigates the use of morphemic division in H497. For the purpose, the corpus instances have been classified in terms of the word-formation process involved, whether prefixation, suffixation or composition. This taxonomy has allowed us to rank the writer's preferences and to determine the type of prefixes and suffixes which are more consistently split.

Examples do not proliferate in the data under scrutiny as they amount to 68 instances in H497, giving the impression that the scribe recurs to morphological division only if there is not enough space at the margin. Otherwise, the words are generally written solid.

3.4.1. Suffixation

This term comprises both derivatives and inflectional endings. There are 32 instances in H497, of which 18 are suffixes and 14 are inflections. The most productive suffix is *-ness* (x11), followed by *-ly* (x5), *-oun* (x1), and *-atyf* (x1), respectively. The most frequent inflection, on the other hand, is obviously the plural marker *-es*/*-en* (x4), followed by *-yng* and *-ed* for present and past participles (x4), the third person inflection *-yth*/*-eth* (x3), infinitival *-e* (x1), the genitive singular morpheme *-es* (x1), and *-yst* (x1) for the absolute superlative, respectively. No matter whether a derivative or an inflection, words are systematically broken before the corresponding suffix/inflection, as in *man-ys*, *strong-yst*, *gret-ly*, *destroy-eth*, *go-ynge*, *heuy-ed*, etc.

Besides, morphological division is often used to avoid the sequence of three vowels in a series, as in *ey-en*. There are, however, some exceptions like *dygesty-oun* where the writer, in spite of using the typical morphemic division (*digest-youn*), purposely moves the boundary of the word to prevent the likely sequence of three vowels in a series.

3.4.2. Prefixation

There are 20 instances of this group in H497, of which *a-* is the most productive (x10), followed by the past participle marker *y-* (x4), *re-* (x2), *be-* (x1), *de-* (x1), *em-* (x1) and *for-* (x1), respectively. Both native and non-native prefixes are found as in *a-way*, *a-slepe*, *y-dronke*, *re-solued*, *de-coccion*, *for-seyd*, *em-plastre*, etc.

3.4.3. Composition

It amounts to 16 instances in the corpus and, in view of our taxonomy, adverbs are especially prone to be broken in these contexts, as in the following cases: *ouer-muche*, *after-warde*, *doun-fallyng*, *ut-myst*, *þer-with*, etc. Unlike other prose compositions of the period where the identity of the constituents is consistently preserved, the writer does not hesitate to break them elsewhere depending on the spatial constraints, such as *overco-meres* or *oftenty-mes*.

3.5. Anomalous divisions

This section presents a small group of instances which cannot be justified in light of the previous rules. There are 14 irregular divisions in H497, a negligible figure if compared with the total number of instances analysed. Given the scribe's masterly use of the phenomenon, these deviations are plausibly the result of the writer's need to fill the entire length of the line, thus avoiding the use of unnecessary blank spaces therein.

Irregular divisions in H497 may be subdivided into three main groups in view of the particularities of the splitting. First, we have found a reduced number of long-stem monosyllables which follow the V - V rule, irrespective of the pronunciation of the digraphs, as in *gre-ef*, *ho-ol*, *sa-am* and *so-or*. The scribe recurs to the use of digraphs in these cases to avoid the stranding of a single consonant in the following line. This is particularly noteworthy in the cases of *ho-ol* and *sa-am* because these words are elsewhere spelt as *hole* and *same* in H497, respectively, leading us to conclude that orthography is here affected by word division. As the adjacent vowels represent a single sound, they may not be broken and the best option would have been to run over the entire word to the following line.

Second, there is also another group of monosyllables presenting an anomalous syllabic division like *my-ght* and *myg-ht*, on the one hand, and *tho-ugh*, *thou-gh*, *thor-ugh*, on the other. In light of the consistent practice of the scribe, some basic principles are accordingly misused here. The general impulse behind word division is to split a word into syllables, so the CV - CV and the C - C rules predominate. In addition to this, the writer is also meticulous not to split letters which form one sound nor a word so that the second part is unpronounceable, aspects which are manifest in the examples at hand. Again, more suitable would have been the general avoidance of word division in the previous monosyllables (see Massey 1934: 402).

Third, there is another group containing disyllabic words such as *capyte-ll* and *me-dle*. As above, the first is anomalous insofar as its second part is unpronounceable and, therefore, the CV - CV rule would have been more appropriate to provide either *capy-tell* or *ca-pytell*, avoiding the separation of the head from its corresponding coda. *Me-dle*, in turn, is more significant because it makes use of the CV - CV rule assuming an erroneous syllabic boundary of the verb **me-dlen*. In need of division, however, the use of the C - C rule seems more convenient in terms of its syllabic boundaries, thus giving a more standard division like *med-le*.

4. Conclusions

In the present paper word-division usage was surveyd in a late Middle English scientific text, which has been purposely chosen on account of the following reasons, one quantitative and the other qualitative. The data analysed amounts to 35,950 running words, providing us with 954 word-division instances, a sizeable input in view of the frequency of the phenomenon if compared with other mediaeval manuscripts, whose use is testimonial. Qualitatively speaking, on the other hand, H497 is written by a sole hand using a legible 15th-century secretary script, represented with the utmost care for the sake of the preciousness of the composition. This fact, together with the consistent spelling practice of the scribe, makes it an appropriate input for the analysis of word division in the period, wherefrom the following conclusions have been drawn:

a) The first objective of this paper was the analysis of scribal attitudes towards word division in order to obtain statistical evidence about the distribution of the two guiding principles, either morphemic or phonological. In H497, the phonological has been found to outnumber the morphological insofar as they occur in 91.4% and 7.12% of the cases, respectively (the anomalous instances represent 1.46% in the corpus). Even though this is the standard practice among late Middle English scribes, there are also pieces which depart from this trend. Hladký affirms that word division in Caxton's time is highly erratic whilst a revival of morphological division takes place among Dryden's contemporaries (Hladký 1985b: 141). For this reason, it is thorny to offer general conclusions about word division in a particular period because its usage ultimately depends on the particular choices and idiosyncrasy of the scribe. Every piece of prose, therefore, should be viewed within its own spectrum, with its exclusive rules of syllabification and word division, which may differ from other similar compositions, irrespective of its chronology.

b) The second objective of this paper was the classification of the corpus instances to determine, on a statistical basis, the orthographic resources available to mediaeval scribes in the vernacular. Figure 1 below presents the statistical distribution of the examples in terms of the phonological rule used. For the sake of comparison, the figures have been normalized to a text of 10,000 words.

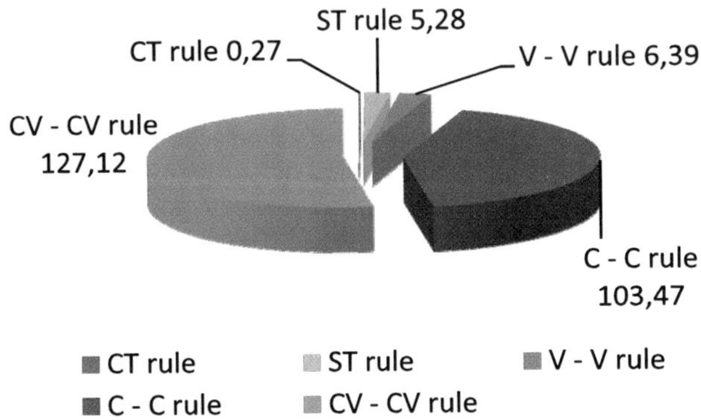

Figure 1. Phonological rules in H497.

As illustrated, the CV- CV rule is the most widespread phonological boundary with 127.12 instances every 10,000 words, followed by the C - C rule which amounts to 103.47 examples. The occurrence of the V - V and the ST rules is practically negligible as they amount to 6.39 and 5.28 instances, respectively. In light of this, it may be concluded that the traditional tenet about the scribal inconsistency of word division needs to be discredited in view of the level of standardization of prose texts like this one. Although there is a consistent practice followed with some level of homogeneity in H497, other prose pieces are found to be more erratic in terms of word division. We will then have to wait until the late modern English period to observe a more conventional pattern of word-division usage in English (Calle-Martín forthcoming).

c) The third objective of this paper was to survey the concept of syllable boundary in Middle English manuscript composition. Notwithstanding the level of scribal variation, it is possible to postulate the concept of syllable boundary for mediaeval scribes in the vernacular, which is primarily based on phonological considerations. A word is therefore considered to contain as many syllables as vowel sounds (with the exception of final *-e* in native words), which the writer proceeds to divide on account of the following variables: the space available at the margin of the folio and the intrinsic nature of the syllable involved. In view of these variables, the writer ultimately decides which strategy to use by using the following continuum: open syllable division → close syllable division → vowel - vowel division → morphological division, besides the contribution of other minor rules like the CL or the ST rules, for instance. Irrespective of the rule used, it is generally avoided to split letters forming one sound or to split a word so that the second part becomes unpronounceable.

In this vein, the modern reader is then in a position to guess the likely breaking points of a Middle English word by using this continuum. A word like *coryaundre*, for instance, would be in principle divided as follows *co-ryaundre* (the CV - CV rule), *coryaund-re* (the C - C rule), and *cory-aundre*/*corya-undre* (the V - V rule), though the first one is more likely. *Lyquorice*, on the other hand, may be broken as *ly-quorice*/*lyquo-ryce* (the CV - CV rule), *lyqu-oryce* (the V - V rule) or *lyquor-ice* (suffixation), where the open syllable division would again predominate.

References

Burchfield, Robert 1994. Line-End Hyphens in the *Ormulum* Manuscript (MS Junius I). In Malcolm Godden, Douglas Gray, and Terry Hoad (eds) *From Anglo-Saxon to Early Middle English*, Oxford: Clarendon Press, 182-187.

Calle-Martín, Javier Forthcoming. A Corpus-based Approach to Line-Final Word-division in Early English Handwriting. In Jacob Thaissen and Hanna Rutkowska (eds) *Scribes, Printers, and the Accidentals of Their Texts*, Frankfurt: Peter Lang.

Calle-Martín, Javier / Miranda-García, Antonio 2005. Editing Middle English Punctuation. The Case of MS Egerton 2622 (ff. 136-152). *International Journal of English Studies*. 5.2, 27-44.

Calle-Martín, Javier / Miranda-García, Antonio 2008. The Punctuation System of Elizabethan Legal Documents: the Case of G.U.L. MS Hunter 3 (S.1.3). *The Review of English Studies*. 59, 356-378.

Cross, Rowin 2004. *A Handlist of Manuscripts Containing English in the Hunterian Collection Glasgow University Library*. Glasgow: Glasgow University Library.

Denholm-Young, Noël 1964. *Handwriting in England and Wales*. Cardiff: University of Wales Press.

Derolez, Albert 2003. *The Palaeography of Gothic Manuscript Books. From the Twelfth to the Early Sixteenth Century*. Cambridge: Cambridge University Press.

Hamel, Christopher de 1992. *Scribes and Illuminators*. London: British Museum Press.

Hector, Leonard Charles 1958. *The Handwriting of English Documents*. London: Edward Arnold.

Hladký, Josef 1985a. Notes on the History of Word Division in English. *Brno Studies in English*. 16: 73-83.

Hladký, Josef 1985b. Word Division in Caxton and Dryden. *Philologica Pragensia*. 28: 135-141.

Hladký, Josef 1987. Word Division and Syllabification in English. *Brno Studies in English*. 17: 123-130.

Johnston, Edward 1906. *Writing and Illuminating, and Lettering*. London: Pitman.

Keiser, George R. 1998. *A Manual of the Writings in Middle English 1050-1500. Volume X. Works of Science and Information*. New Haven, Connecticut: The Connecticut Academy of Art and Sciences.

Kurath, Hans / Kuhn, Sherman M. / Reidy, John / Lewis, Robert E. (eds) 1956-2001. *Middle English Dictionary*. Ann Arbour: University of Michigan Press. [Electronic dictionary: <http://quod.lib.umich.edu/m/med/lookup.html>].

Lutz, Angelika 1986. The Syllabic Basis of Word Division in Old English Manuscripts. *English Studies*. 67.3: 193-210.

Massey, B.W.A. 1934. The Division of Words. *Modern Language Notes*. 49.6: 400-404.

Osley, A.S. 1980. *Scribes and Sources. Handbook of the Chancery Hand in the Sixteenth Century*. London: Faber and Faber.

Parkes, Malcolm B. 1992. *Pause and Effect. An Introduction to the History of Punctuation in the West*. Hants: Scolar Press.

Petti, Anthony G. 1977. *English Literary Hands from Chaucer to Dryden*. Cambridge, Mass: Harvard University Press.

Powell, Marcy S. 1984. Word Division. *English Studies*. 65.5: 452-458.

Roberts, Jane 2005. *Guide to Scripts Used in English Writings Up to 1500*. London: the British Library.

Scott, Mike 2008. *WordSmith Tools 5.0*. Oxford: Oxford University Press.

Tannenbaum, Samuel A. 1930. *The Handwriting of the Renaissance*. New York: Columbia University Press.

Young, John / Aitken, P. Henderson 1908. *A Catalogue of the Manuscripts in the Library of the Hunterian Museum in the University of Glasgow*. Glasgow: James Maclehose and Sons.

MARK CHAMBERS

What is this, a betell, or a batowe, or a buskyn lacyd?: Lexicological Confusion in Medieval Clothing Culture

The title of this paper is a quote from a scene from John Skelton's play of *Magnyfycence*,[1] probably written around 1519, and published by William Rastell around 1530.[2] In the quote, a two-faced vice named Cloked Colusyon takes issue with another character, a fellow vice who has just entered the stage. Courtly Abusyon, as he is called, is dressed in the latest French fashion, and he struts about the acting space, showing his attire. Cloked Colusyon first assumes he is French, asking him, "De que pays este vous?" (l. 748). Cloked Colusyon then takes particular issue with Courtly Abusyon's shoes, asking him, "By God, syr, what nede all this waste? / What is this, a betell, or a batowe, or a buskyn lacyd?" (ll. 753-4).

Typical of the medieval allegorical drama, the vices of the play propel the action through deception and duplicity, by assuming false names and costumes in order to hoodwink the protagonist. Coked Colusyon's upbraiding of Courtly Abusyon's dress plays on the "conventional association between courtliness, Frenchness and vice" (Walker 2000: 350),[3] and he elicits humour by pointing out the

1 John Skelton, *Magnyfycence* [c. 1519] has been edited by Greg Walker (2000: 368, l. 754), and by Peter Happé (1979: 211-311). References to the play will be from Walker's edition and will be cited by line number(s) in parentheses following the quotation.

2 For a discussion of the play's auspices, see Walker's introduction (2000: 349-351), and Happé's introduction (1979: 39-41).

3 Walker cites a commentator on Henry VIII's expulsion of the 'minions' in 1519-20: "'they [the minions] were all French, in eatng, drynkyng and apparel, yea and in French vices and bragges, so that all the estates of

absurdity of his overly-long shoes. As Leigh Winser points out, his metaphor is nothing new. A few years earlier, in Pierre Gringore's *Sottie des chroniqueurs* of 1515, we find a reference to "newfangled *gorriers* [ ] / Wearing slippers very large, / Fashioned like a boat or barge"[4] – close enough, in fact, for Winser to suggest that the *Sottie* must have influenced Skelton's play (1981: 85-6).

Beyond its literary and social background, however, Cloked Colusyon's fashion criticism draws attention to the concept of *naming*, especially as it relates to the commodities of fashion – of clothing and textiles. Readers of the play – including editors, researchers, actors, students, enthusiasts, and so on – must imagine what Courtly Abusyon's large shoes must have looked like, based on the not-inconsiderable graphic and archaeological evidence of early fifteenth-century footwear, but also based on the words Cloked Colusyon uses to describe the shoes. They are similar to a beetle (a beating implement), some sort of boat, or a buskin; they flap about, we are told, as Courtly Abusyon "fotys… lyke a swanne" (l. 764).

While there are numerous dictionaries, histories and encyclopaedias available, the accurate union of *name* and *thing* with regard to clothing, textiles and fashions are a regular source of confusion for commentators on medieval history and literature. In the 1930s, the costume historian Mary Houston was forced to concede that, "[o]ne subject of importance is, unfortunately, also a source of confusion, namely, the nomenclature of medieval garments. There is much diversity of opinion" (1996: vi). In their more recent study of medieval dress, Perrine Mane and François Piponnier qualify this assertion:

> it remains difficult to relate the written word to the *realia* supplied by archaeology or the images of iconography. Latin and the vernacular languages use different names to designate the same object and regional variants accompany fabrics, fur and fashions across national boundaries.(1997: 12)

Englonde were by them laughed at,'" (Hall 1809: 598; cited in Walker 2000: 350). See also Scattergood (2000: 64-5).

4 Picot (1902-1912: II, 11.110-12); cited and trans. by Winser (1981: 85-6).

Most recently, in her study *Fashion in Medieval France*, Sarah-Grace Heller offers this assessment:

> the treatment of fashion based on visual representations and artifacts is obliged to analyze historic costume and textiles with the aid of modern technical vocabulary. Art history and the philology of apparel remain separate because there exist no medieval fashion manuals that juxtapose words and images, an anachronistic expectation motivated by the existence of such fashion manuals today. (2207: 48)

This final statement predicates two things, however, both of which Heller goes on to admit:

1. that however "anachronistic" it may prove, a thorough consideration of textual evidence alongside (or parallel) to graphic and archaeological evidence is largely wanting, leading to continued repetition of past assumptions and errors based on limited-disciplined studies; and

2. that if one wishes to begin to truly understand something tangible but largely absent – such as medieval clothing – one must consider elements from all extant evidence: textual, graphic, and archaeological. One must be careful, however, to keep these evidences distinct, and to avoid drawing undue assumptions by comparison.

Part of the obvious problem is that *names* for clothing – and for fashion in particular – are borrowed, adapted and discarded as quickly as the items and practices themselves. With regard to elongated shoes, for example, we might take a look at the well-established history of the late-fourteenth and fifteenth-century "piked" variety, often described or shown worn by fashionable young men.

1. Various forms of Evidence – A Test Case

A mid fifteenth-century miniature from Brussels (Biblioteque Royale 8, folio 7, from David Aubert's *Histoire de Charles Martel*) depicts a scene in which Charles the Bold is shown "surprising" David Aubert, who is busy working on a manuscript.[5] In the scene, Charles' courtiers wear many of the fashions associated with aristocratic men's dress in Northern Europe through the 1450s and 1460s – elongated shoes, exceptionally shortened tunics with "bag-pipe" sleeves[6] and a variety of trendy hats.[7] The "piked" variety of shoes visible on the young men to the far left and right of the miniature are often referred to as *poulaines*, and they were particularly popular in the middle of the fifteenth century, when we first encounter the word in British, as well as Continental contexts.[8] But the fashion of wearing elongated points

---

5 Reproduced in Kren and McKendrick (2003: 231-233, no. 55).

6 Fifteenth-century "bagpipe" sleeves are often referred to as *pokes* in Middle English, cf. "poke *n.*," *Middle English Dictionary* [*MED*] 1998: *Middle English Compendium* online, 1 Jan. 2008, <http://quod.lib.umich.edu/cgi/m/mec/med-idx?size=First+100&type=headword&q1=poke&rgxp=constrained>. For a discussion of the fashion, see Houston (1996 [1939]: 159, illust. fig. 307, p. 176); Norris, *Medieval Costume and Fashion* (1999 [1927]: 361, illust. fig. 498); and Phillis Cunnington, *Medieval & Tudor Costume* (1968: 28). Houston (252) notes that the fashion saw its initial popularity in England during the reign of Richard II. By the first decade of the fifteenth century we find a reference to a "maxime togarum cum profundis et latis manicis, uocatis uulgariter pokys, ad modum unius bagpipe formatis" [large gown with full and broad *manicas*, called in the vernacular pokys, made in the fashion of bagpipes], in the *Historia Vitae et Regni Ricardi Secundi*, ed. G. B. Stow, Jr. (1977: 168, my translation).

7 Many kinds of fashionable hats enter men's fashion during the mid fifteenth century, including what have since been described as 'bowler' and so-called 'turkey bonnet' styles shown in the miniature; cf. Norris, *Medieval Costume and Fashion* (1999 [1927]: 433-437); Hou ston (1996 [1939]: 173-5); and Cunnington and Cunnington (1973: 147-9). For a discussion of medieval headwear in English funerary monuments, see Gardner (1950: 4-13).

8 For the word's use in English, the *Oxford English Dictionary* suggests a development from, "Anglo-Norman *poleine* (1464-9 or earlier ... and Middle French, French (now hist.) *poulaine* (1375-9 as *poullaine*, denoting the

on shoes in England was popular before the fifteenth-century Continental word *poulaine* was adopted or adapted to describe the practice.

If we look further back in the English drama, in the early fifteenth century we find a reference to a type of elongated, pointed shoe that pre-dates the *poulaine* that had been adorning the feet of English noblemen for a few decades. In the play *The Castle of Perseverance*, the lowly protagonist Humanum Genus (Mankind) signals his movement from innocence to sin at the beginning of the play by walking from his position in the central castle over to the scaffold of Mundus (the World).[9] Mundus welcomes Humanum Genus up to his scaffold, and he offers him a glorious new livery, decorated with "bysawntys bryth" (l. 588) – gold ornaments or pattern, originally called after Byzantine coins.[10] Then, Humanum Genus welcomes each of the seven deadly sins up to the scaffold to join him. Superbia (Pride) instructs him to wear long "crakows,": "vse these new jettys: Loke þou blowe mekyl bost / Wyth longe crakows on þi schos" (ll. 1058-9).

In English, the word *crakow* first referred to the point itself and then to the entire shoe, and was accompanied by a verbal form,

---

pointed toe; *c*1386-9 or earlier as *polayne*, denoting the shoe ... the shoes and their characteristic long toe being so called on account of their supposed Polish origin...," "poulaine, *n.*," *The Oxford English Dictionary* [*OED*], 2nd ed. 1989: *OED Online*, 1 Jan. 2008, <http://dictionary.oed.com/cgi/entry/50185612>.

9 *The Castle of Perseverance* has been edited by Mark Eccles for the EETS (1969: ll. 574-614). Further references to the play will be from this edition and will be cited by line number(s) in parentheses following the quotation.

10 The *MED* defines *besaunt* as: 1) "A gold coin of Byzantium; any of several similar coins minted in Western Europe; also, a silver coin used in Western Europe…;" and 2) "A bezant used as an ornament, an ornament resembling a bezant," *MED, Middle English Compendium* online, 1 Jan. 2008 <http://quod.lib.umich.edu/cgi/m/mec/medidx?size=First+100&type=headword&q1=besaunt&rgxp=constrained>. For a discussion of "bezants" used as ornamentation in clothing (particularly in the mid-fourteenth century), see also Newton (1999 [1980]: 25, 34, 36, 43, 80 and 98).

crakowen, to provide with points.[11] The conventional history of the *crakow* shoe usually attributes the name and fashion to the court of Richard II's wife, Anne of Bohemia (daughter of the Holy Roman Emperor Charles IV, on whose border the important trading city Krakow, in Poland, stood),[12] and this etymology has been handed down. In the 20th century, we find it repeated in the influential medieval fashion manuals by Norris and by Houston, both of whom suggest that "crackows" made their way into Britain on the feet of Queen Anne's courtiers (Houston 1996 [1939]: 210; Norris, *Medieval Costume and Fashion* (1999 [1927]: 253). In her history of *Joan of Arc*, Marina Warner conflates this etymology with that of French *poulaines*: "'Poulaines', the later, English winkle-pickers, first came into vogue after Anne of Bohemia's suite brought this 'Polish' fashion to the English court" (Warner 1981: iii.170).

But "crackows" and similarly pointy shoes appeared in England before Richard married Anne in 1382 – in fact, even before her birth in 1367.[13] In the mid 1360s, John of Reading had ridiculed the

---

11 "crakou, *n.*," *MED* 1998: *Middle English Compendium* online, 1 Jan. 2008 <http://quod.lib.umich.edu/cgi/m/mec/med-idx?type=id&id=MED10180>; "crakouen, *v.*," *MED* 1998: *Middle English Compendium* online, 1 Jan. 2008 <http://quod.lib.umich.edu/cgi/m/mec/med-idx?type=id&id=MED10181>; and "crakow, *n.*," *OED*, 2nd ed. 1989: *OED Online*, 1 Jan. 2008 <http://dictionary.oed.com/cgi/entry/50053114>.

12 Many contemporary historians take their derivation from a reference to the year 1394 in a *Histoira Vitae et Regni Ricardi II*, attributed to a monk of Evesham, who declares:
Cum ista Regina venit [venerunt] de Boëmia in Angliam abusiones illae execrabiles, sotulares sil. [scilicet] cum longis rostris (Anglice *Cracowys* vel *Pykys*) dimidiam virgam largiter habentes, ita ut oporteret eos ad tibiam ligari cum cathenis argenteis, antequam cum eis possent incedere.
[With this queen there came from Bohemia into England those accursed vices, namely shoes with long beaks (English *Cracowys* or *Pykys*) half a yard in length, thus it was necessary for them to be tied to the shin with chains of silver before they could walk with them.] Edited by T. Hearne (1729: 126, my translation).

13 If its suggestion for the year 1362 is accurate, the late fourteenth-century *Eulogium Historiarum* attributed to Thomas of Malmsebury precludes the etymology through Anne:

"sotularibus... lateraliter rostratis," ("shoes... with sideways points/beaks"), worn by many noblemen,[14] and an anonymous poem of 1380 criticises the "rostra fovent ocrearum" ("points of the boots/leggings they wear/promote").[15] When uniting objects, names and etymology, it is often difficult to localise and temporalize contemporary practices. The name "crackow" (as with "poulaine") obviously has its source in Poland, but the fashion itself did not necessarily make its way into Britain newly formed, on the feet of a Bohemian princess' courtiers, as Warner and others continue to suggest.

Elongated footwear and points-on-shoes are part of what fashion historians have called the "logic of excess and exaggeration," where incremental changes are made to a standardised silhouette based on the natural human form (Heller 2007: 9). Indeed, exaggerations such as these are often the most distinctive visual markers of change in fashion history, when trendsetters raise or lower a neckline, exaggerate the sleeve or shoulder, shorten a tunic or broaden a ruff. These changes are usually cumulative rather than dramatic, until an extreme is reached, at which the fashionable move towards greater austerity (one might think of "flared" trousers in the 1960s and 70s, of platform shoes, of large shoulder-pads in the 1980s and early 90s, and of the reactions to such fashions in later generations).

But as Heller postulates, "fashion is paradoxically characterized by both constant change and relative stasis...like wisteria, which blooms on old wood...." (2007: 30). Pointy shoes were nothing new in Britain with the advent of the so-called "crackow" in the fourteenth century, just as they weren't in Skelton's play in the early sixteenth. We find descriptions of stuffed points on men's shoes in England as

---

Habient etiam sotulares rostratas in unius digiti longitudine quae *crakowes* vocantur; potius judicantur ungular...daemonum quam ornamenta hominum. Edited by F. S. Haydon (1858-63: vol. III, 231).

14 The *Chronica Johannis de Reading et Anonymi Cantuariensis* is edited by J. Tait (1914: 167).

15 British Library, MS Harley 941, fols. 21v-23v (1450-1500) l. 142, transcribed by Richard Firth Green (1991: 331). Green also discusses the dating in his article (330-341)

far back as the reign of William II (Heller 2007: 55). In his early twelfth-century *Historia Ecclesiastica*, Orderic Vitalis complains about the trivial and debased fashions from the late eleventh century, and we get glimmers of a stuffed-towed, elongated shoe favoured by some courtiers in northern France:

> [Of Count Fulk[16]:] Hic in multis reprehensibilis et infamis erat. multisque uitiorum pestibus obsecundabat. Ipse nimirum quia pedes habebat deformes, instituit sibi fieri longos et in summitate acutissimos subtolares. Ita ut operiret pedes, et eorum celaret tubera quæ uulgo uocantur unions. Insolitus inde mos in occiduum orbem processit. Leuibusque et nouitatum amatoribus uehementer placuit. Vnde sutores in calciamentis quasi caudas scorpionum quas uulgo pigacias appellant faciunt. idque genus calciamenti pene cuncti diuites et egeni nimium expetunt.
>
> [Count Fulk was a man with many reprehensible, even scandalous, habits, and gave way to many pestilential vices. Being a man with deformed feet he had shoes made with very long and pointed toes, to hide the shape of his feet and conceal the growths that are commonly called bunions. This encouraged a new fashion in the western regions, which delighted frivolous men in search of novelties. To meet it cobblers fashioned shoes like scorpions' tails, which are commonly called 'pulley-shoes', and almost all, rich and poor alike, now demand shoes of this kind.] (M. Chibnall, ed. and trans. 1973: VIII.10.186-7)

Orderic then relates how this fashion spread to Norman England:

> Robertus quidam nebelo in curia Rufi Regis[17] prolixas pigacias primus cepit implore stuppis. et hinc inde contorquere instar cornu arietis. Ob hoc ipse cornardus cognominatus est. cuius friuolam adinuentionem magna pars nobelium ceu quoddam insigne probitatis et uitutis opus mox secuta est. ...Femineam mollitiem petulans iuuentus amplectitur. Feminishque uiri curiales in omni lasciuia summopere adulantur. Pedum articulis ubi finis est corporis colubrinarum similitudinem caudarum imponunt. quas uelut scorpions præ oculis suis prospiciunt.
>
> [Robert, a certain worthless fellow at King Rufus's court, first began to stuff the long 'pulley-toes' and in this way bend them into the shape of a ram's horn. As a result he was nicknamed Cornard. The frivolous fashion he had set was soon imitated by a great part of the nobility as if it had been an achievement of great worth and importance. ...Our wanton youth is sunk in effeminacy, and courtiers, fawning, seek the favours of women with every

---

16 Fulk le Rechin, Count of Anjou (1068-1109).
17 William Rufus (William II, reigned 1087-1100).

> kind of lewdness. They add excrescences like serpents' tails to the tips of their toes where the body ends, and gaze with admiration on these scorpion-like tails.] (Chibnall, ed. and trans. 1973: VIII.10187-9)

Here we have a detailed description of the shoes ("calciamentis quasi caudas scorpionum"), later "stuffed" and bent into the shape of a ram's horn ("instar cornu arietis"); we have a class of wearers ("diuites et egeni" in Orderic's time, "magna pars nobelium" in Rufus' court); we have a name ("pigaces")[18] and even a suggestion towards historical or etymological development (through a certain "Robert"). While we must take the descriptive accuracy of Orderic' diatribe with a sizeable pinch of salt, it is clear that stuffed, elongated toes on shoes in a manner possibly very similar to later *crakows* and *poulaines* were being worn in England as early as the late eleventh century.[19] In other words, Superbia's *crakows* and Courtly Abusyon's conspicuous waste as developments of a mutable, but common sartorial theme, rather than a fashion trend newly and wholly launched out of Eastern Europe in the late fourteenth century.[20]

If we look at the major handbooks on the medieval clothing and fashion published in the last century – by F. H. Fairholt (1909), Norris (1926, 1927, 1999 [1927]), Houston (1996 [1939]), the Cunningtons (1973 [1952]), Francis Kelly and Randolph Schwabe (1968), Newton

---

18 The word comes from Old French (cf. "pigace, *n.*," *MED* 1998: *Middle English Compendium* online, 1 Jan. 2008 < http://quod.lib.umich.edu/cgi/m/mec/med-idx?type=byte&byte=142312213>), although there are no English attestations with this specific sense. The *Anglo-Norman Dictionary* [*AND*] offers one definition of "a long, pointed shoe," with an attestation from the second half of the thirteenth century, cf. "pigace, *n.*," *AND*, 2nd ed., W. Rothwell *et al.*, eds. 2005: *The Anglo-Norman On-line Hub*, Publications of the Modern Humanities Research Association, 17 (I), 1 Jan. 2008, <http://www.anglo-norman.net/gate/>. Further references to the *AND* online will be from this date and site.

19 For a brief discussion of so-called "pigaces", see also Norris, *Medieval Costume and Fashion* (1999 [1927]: 58-9 and figs. 75 & 76).

20 In fact, in a footnote in her edition of Orderic's *Ecclesiastical History*, Marjorie Chibnall breezes over any distinction between Fulk's *pigaces* and the later *poulaine*: "These shoes, which persisted in France into the seventeenth century, were known as 'souliers à la poulaine'. A specimen may be seen in the Musée de Cluny in Paris." (1973: 187n4).

(1999 [1980]), Manes and Pipponier (1997), Margaret Scott (1980, 2007) and others[21] – we sometimes find a distinct reliance on the graphic and archaeological evidence, often to the detriment of literary texts and account records which can often bolster understanding of look and feel. Heller argues, for example, that the commonly accepted idea that fashion itself somehow "began" with the shortened surcoat fourteenth-century – particularly Burgundy (an idea first suggested by Paul Post) – is a product of this reliance on graphic evidence and the greater availability of illuminations and other visual representations from this period than earlier periods (2007: 48-54).[22]

2. Dictionary Evidence

Conversely, by their very nature, dictionaries are primarily limited to textual evidence. Moreover, they are usually monolingual, and – according to lexicographers such as William Rothwell, David Trotter, Laura Wright and others – can inadvertently give a false impression of the multilingual realities of medieval Britain (Rothwell: 1993, 1994; Trotter 2007: 73-90; Wright 1996). Mixed-language texts ("macaronic", to use Wright's term) are often far more plentiful than linguistically homogeneous ones, particularly after the Conquest: "it is easier today to find macaronic documents from the late Medieval period in the Record Offices than it is to find monolingual texts, which is a testament to both the number that were written and their contemporary importance" (1996: 6). Linguistic boundaries – particularly with technical writing – were certainly not as distinct as they might be today. Wright puts it more strongly: "I contend that we

21 This list only represents a few major handbooks on medieval fashion and dress and is not intended to be comprehensive.

22 Heller makes reference to Paul Post's "Die französische-niederländische Männertracht einshchliesslich der Ritterrüstung im Zeitalter der Spätgotik, 1350-1475 (1910); and "La naissance du costume masculine modern au XIVe siècle," (1952).

have been strait-jacketed by our own monolingual perspective, and our social and political desire to find linguistic and hence cultural purity" (1996: 7).

While the dictionaries often carry attestations that demonstrate mixed-language writing and occasional "single lexeme switches" – where a word with an etymon from one language is inserted into a text of another, for whatever reason – they can represent an attempt to codify or circumscribe a single linguistic boundary at the expense of contemporary, "macaronic" linguistic realities. Moreover, a complete reliance on the Dictionaries, or else an isolated examination of the visual evidence, tends to leave the picture half painted – the cloth half fulled, as it where. One cannot get to the bottom of Courtly Abusyon's footwear by considering the dictionary evidence alone – especially if that consideration fails to consider the multi-linguistic realities of medieval Britain. Conversely, a consideration of solely graphic or archaeological evidence is unable to realise accurate naming, leading to, among other things, misapplied references to the textual evidence and a continued compounding of past errors and undue assumptions about medieval cloth and clothing culture.

3. The Lexis of Cloth and Clothing Project

These are precisely the sort of bounded approaches to the study of medieval clothing that a new, AHRC-funded Project hopes to address. Entitled "The lexis of cloth and clothing in Britain c. 700-1450: origins, identification, contexts and change," the five-year project (launched early in 2007) will endeavor to create a searchable database of lexical, graphic and archaeological evidence of medieval cloth and clothing. Its basis will be the medieval lexis, including all of the relevant languages of the British Isles (including Anglo-Norman, Cornish, Flemish, Irish, Medieval Latin, Manx, Middle English, Norn, Norse, Old English, Old Scots, Scots Gaelic and Welsh). It is hoped that by assembling the often polyglot linguistic evidence for medieval cloth and clothing of the British Isles, alongside appropriate graphic and archeological evidence, a greater understanding of garments, fabrics and their proper names will be achieved and an exceptionally useful resource will be made available to scholars of clothing history, material culture and historical linguistics.[23]

No doubt the project's initial remit to begin assembling the *lexis* of medieval cloth and clothing will raise a number of issues, as we begin by prioritizing the textual evidence. To paraphrase Roland Barthes' *Système de la Mode*, "it is not the object but the name that inspires desire."[24] It is reasonable (according to Barthes) to proceed from the word to the real when discussing fashion in particular, for fashion and the desire for material consumption is a creation of language. But, although it begins by assembling a lexis, the Project's overall aims are "cross-disciplinary", as subsequent stages will see Project members playing to their own strengths, adding to the larger corpus of data about names for medieval cloth and clothing that builds on the textual evidence, rather than being limited to it.

4. Clothing Toponyms and the Dictionaries: *ermine* and *rencian*

To extend the example of *crackows* and *poulaines* further, we can see how misattributed toponymns are not necessarily isolated incidents in medieval clothing terminology. The raw material "ermine" (Spanish *armiño*), is, of course, a luxurious staple of the medieval wardrobe. Edward III reserved its use for the royal family and nobles worth a thousand pounds *per annum* or more (Norris, H. 1999 [1927]: 283), its semblance in heraldic patterns is well attested, and one would be hard-pressed to find a medieval romance where expensive ermine isn't

23 Further information can be found at the project's website: <http://lexisproject.arts.manchester.ac.uk/>.

24 "Ce n'est pas l'objet, c'est le nom qui fait desire," (Barthes 1967: 9); cited and translated by Heller (2007: 35).

lining or trimming an elegant garment – as in *Kyng Alisaunder*: "cloþen hem wiþ grys & *ermyne*," (G. V. Smithers 1952: l. 4970, my emphasis), or *Sir Gawayn and þe Grene Knyȝt*: "a meré mantyle.../fayre furred wythinne with fellez of þe best, / Alle of *ermyn* in erde, his hode of þe same;" (N. Davis, ed. 1967: 25, ll. 878-81, my emphasis).

The word itself developed in Middle English from Old French with both nominal and adjectival forms, there is a tradition linking the word to *Armenian* and *Armenia* in the Romance languages, based on a supposed place of origin. The *OED* traces this association as far back as Julius Pollux in around 180 A.D.[25] It also points out that this etymology was quite popular in late-medieval Europe, and we often find the word representing modern ermine (the fur) or modern Armenian. Both the *Altfranzösisches Wörterbuch* and the *AND* have attestations for both senses.[26] It does not seem to have been as prevalent in Middle English, but we do find homographic forms, such as that in Chaucer's Monk's description of the fearsomeness of the widowed Zenobia:

> The Emperour of Rome, Claudius...,
> Ne noon *Ermyn*, ne noon Egipicien,
> Ne Surrien, ne noon Arabyen,
> Withinne the feeld that dorste with hire fighte....[27]
> *MK*, VII.2335-40 (Benson, ed. 1987: 246, my emphasis)

The *MED* cites the two senses for substantive forms as two separate entries, but both having the exact same orthography.[28] We do not find

25 "ermine, *n.*," *OED*, 2nd ed. 1989: *OED Online*, 1 Jan. 2008 <http://dictionary.oed.com/cgi/entry/50077592>.

26 Cf. "ermin, hermin, *adj. u. subst.*" in the *Altfranzösisches Wörterbuch* (Tobler and Lommatzsch, eds. 1925-36; 1954-), and "ermin, *s. and a.*" in the *AND*.

27 Geoffrey Chaucer, *The Canterbury Tales*, The Riverside Chaucer, ed. Larry D. Benson, 3rd ed. (Oxford: OUP) *MK*, VII.2335-40, p. 246.

28 "Ermin, *n. and adj.*," *MED* 1998: *Middle English Compendium* online, 1 Jan. 2008 <http://quod.lib.umich.edu/cgi/m/mec/med-idx?type=id&id=MED14343> and "ermin, *n.*," *MED* 1998: *Middle English Compendium* online, 1 Jan. 2008 <http://quod.lib.umich.edu/cgi/m/mec/med-idx?type= id&id=MED14344>.

any explicit reference to this assumed etymology in Middle English, nor do we see any of this confusion in Medieval British Latin *ermina* and its variants (*erminium*, *erminus* [*herm-*] and the heraldic *erminalis*) (cf. "ermina," Latham *et al.*, eds. 1975-: III.795). But the use of the word *ermine* and its variants for both senses in the vernaculars is certainly intriguing. While the vernacular languages on the Continent suggest a tradition of associating *ermine* with Armenia, showing a desire to assign names based on supposed places of origin (as with *crackows* in England), a comparison of the dictionaries shows that this tradition doesn't seem to be attested in the late medieval Latin and is residual, rather than specified, in Middle English attestations of the word and its variant forms.

In an article for *InTRAlinea*, Patrick Leech (1999) conducts a study of "textile terminology" in the *OED*, assembling a wealth of data about specific names for cloth in English usage from about 1000 to 1960. His study is useful in that it focuses "on the historical development of language in a specific field" (textiles), without simply employing technical terms from lexemes in current English. This allows him to focus on diachronic change of the terminology of a particular technical field and to draw conclusions about that terminology's development into modern English.

However, even from a solely lexicological standpoint, complete reliance on a single, monolingual source of data such as the *OED* is bound to lead to misconceptions, something that Leech is forced to admit:

> The need to describe lexical innovation through contact with and contamination from other languages will become apparent when investigating a lexical field such as that of textiles, one that involves large numbers of objects crossing national boundaries as commodities.

Recommending his study as simply "a basis for further research in this field," Leech goes on to analyse the *OED*'s dating and etymological data for some 383 terms denoting natural woven fabrics, of which 100 he lists as toponyms. But his conclusions suffer from the limits of the *OED's* evidence and from the prescriptive nature of such a systematic exercise.

For example, Leech's study misses the *OED's* entry for *rencian*, a word which appears twice in the thirteenth-century in the second part of the Oxford MS Jesus College 29 (Brown 1932: 106; Morris 1988 [1872]: 70). It is described by the *OED* as "Some kind of cloth."[29] A quick comparison with the *MED*, however, demonstrates how the dictionaries themselves do not always agree on etymology. While the *OED* describes the history of *rencian* as "obscure", the *MED* suggests a derivation from Old French *renciien* ("from Reims"), purporting that the word is in fact a toponym, meaning, "A kind of cloth made in or associated with Reims."[30] The *MED*'s notion of a toponym is quite intriguing and provides further information about the development of the word into English – if not of the textile itself. The project will allow data and assessment of the various terms for cloth and clothing from the relevant dictionaries to be compared, analysed and refined. Ultimately, discrepancies between the various dictionaries will be examined and addressed – in essence, pulling together the various philological resources towards the study of a specific technical lexis in a manner in which Leech's article suggests.

5. Contemporary Confusion: Ipswich's "White of Green"

As Cloked Colusyon's joking question about Courtly Abusyon's shoes makes clear, lexical confusion in cloth and clothing culture can affect individual attestations and can skew an easy dictionary entry. Not infrequently do we find contemporary users in some confusion about the proper name used to describe a medieval cloth or garment. In *The Domus Day of Gippeswiche* of about 1436, for example, a scribe apparently mistranslates the Anglo-Norman phrase "*blankes de veyr*" in his Middle English version of *Le Domesday de Gippewyz* (*The*

29 "rencian, *n.*" *OED*, 2nd ed. 1989: *OED Online*, 1 Jan. 2008.

30 "rencian, *n.*," *MED* 1998: *Middle English Compendium* online, 1 Jan. 2008 <http://quod.lib.umich.edu/cgi/m/mec/med-idx?type=byte&byte= 159439200>.

Ipswich Domesday, T. Twiss, ed. 1873). The reference is presumably describing a blanket or woolen cloth of vair or fur (possibly squirrel fur). However, the Middle English translator has offered the rather elusive "white of green," assuming the French original to signify "*blanc de vert*":

> Item de chescun millier de blaunkes de veyr, ij. *s*.; de demi millier xij.*d*. [*The translation:*] Also of eche ml. of qwit of grene, ij.*s*. ; of half a ml., xij.*d*. (Twiss 1873: 190)

This English phrase is unique, and the *MED* suggests the mistranslation.[31] In this case, the original user has created a virtual mis-calqued hapax legomenon, and we only guess what he had in mind when he did so. It stands that this use of the Middle English phrase *qwit of grene* should be included in the project's database (perhaps along with the noun/adjective *whit*, *qwit* = *white*), as this is how the scribe manages to render the French *blaunkes de veyr*, items of cloth from a late medieval Norwich account.

6. Conclusion

As this couple of examples attest, there are many ways in which the accurate names for medieval cloth and clothing have been separated from the actual things they must have represented, at least to contemporary, medieval understanding. An undue reliance on literary evidence, monolingual dictionaries or similarly limited textual sources – or else a sole reliance on graphic and/or archaeological evidence, to the exclusion of the textual – has led many past scholars to undue assumptions. Often, this has led to misattributed descriptions and to the compounding of lexicological errors in the study of medieval cloth and clothing.

---

31 "whit, *n*.," *MED* 1998: *Middle English Compendium* online, 1 Jan. 2008 <http://quod.lib.umich.edu/cgi/m/mec/med-idx?type=id&id=MED52609>.

The AHRC-funded Lexis of Cloth and Clothing in Britain project will seek to address many of these issues, by considering evidence for the lexis from all of the relevant sources of lexicological data for all of the languages of medieval Britain, alongside recent and reliable archaeological evidence and appropriate visual representation from medieval painting, illumination, sculpture, glass-work and so forth. In its final stages, the project will initiate a searchable database of its findings, innovatively designed and illustrated and publicly accessible to interested scholars.

Ultimately, it is hoped that we will be able to more accurately distinguish Courtly Abusyon's "betell,…batowe,…or…buskyn lacyd" for our own generation, as we attempt to shed new light on the medieval cloth and clothing lexis.

References

The Anglo-Norman Dictionary [*AND*], 2nd ed., W. Rothwell *et al.*, eds. 2005. *The Anglo-Norman On-line Hub*. Publications of the Modern Humanities Research Association, 17 (I). 1 Jan. 2008. <http://www.anglo-norman.net/gate/>.

Barthes, R. 1967. *Système de la Mode*. Paris: Seuil.

Beard, C., C. W. Cunnington and P. Cunnington 1960. *A Dictionary of English Costume*. London: Black.

Benson, L. D., ed. 1987. *The Riverside Chaucer*, Oxford: Oxford University Press.

Chibnall, M. 1973. *The Ecclesiastical History of Orderic Vitalis*, Vol. 4, Books VII and VIII. Oxford: Clarendon.

Cunnington, P. 1968. *Medieval & Tudor Costume*. London: Faber.

Cunnington, C. W. and P. Cunnington 1973 [1952]. *Handbook of English Mediaeval Costume*, illust. B. Phillipson and C. Lucas. London: Faber.

Davis, N., ed. 1967. *Sir Gawayn and þe Grene Knyȝt*, 2nd Ed. Origin. ed. by E. V. Gordon and J. R. R. Tolkein, 1925. Oxford: Clarendon.

Eccles, M., ed. 1969. *The Macro Plays* [EETS]. London: Oxford University Press.

Fairholt, F. H. 1909. *Costume in England: A History of Dress to the End of the Eighteenth Century*, 4th ed., ed. H. A. Dillon, 2 vols., London: George Bell and Sons.

Forester, T., ed. and trans. 1854. *The Ecclesiastical History of England and Normandy by Ordericus Vitalis*, Vol. 2. London: H. G. Bohn.

Gardner, A. 1950: Hair and Head-Dress, 1050-1600. *Journal of the British Archaeological Association*, 3rd series, VIII: 4-13.

Green, R. F. 1991. Jack Philipot, John of Gaunt, and a Poem of 1380, *Speculum* 66: 330-341.

Hall, E. 1809. *The Union of the Two Noble and Illustrious Houses of Lancaster and York*, ed. H. Ellis. London.

Happé, P. 1979. *Four Morality Plays*. Harmondsworth, Middlesex: Penguin.

Haydon, F. S., ed. 1858-63. *Eulogium Historiarum sive Temporis*, Rerum Britannicarum medii aevi scriptoris 9. London: Longman, Brown, Green, Longmans and Roberts.

Hearne, T., ed. 1729. Histoira Vitae et Regni Ricardi II, Angliae Regis. Oxford.

Heller, S.-G. 2007. *Fashion in Medieval France*. Cambridge: D.S. Brewer.

Houston, M. G. 1996 [1939]. *Medieval Costume in England and France: The 13th, 14th and 15th Centuries*. New York: Dover.

Kelly, F. M. and R. Schwabe 1968. *A Short History of Costume and Armour, Chiefly in England,* Vol. 1: 1066-1485. London and New York: Benjamin Bloom, 1968).

Kren, T. and S. McKendricks 2003. *Illuminating the Renaissance: The Triumph of Flemish Manuscript Painting in Europe*. Los Angeles / London: J. Paul Getty Museum / Royal Academy of Arts.

Latham, R. E. *et al.*, eds. 1975-. *Dictionary of Medieval Latin from British Sources*. Oxford: Oxford University Press.

Leech, P. 1999. Who Says Manchester Says Cotton: Textile Terminology in the *Oxford English Dictionary* (1000-1960), *InTRAlinea* 2. 1 Jan. 2008. <http://www.intralinea.it/volumes/eng_more.php?id=22_0_2_0_M26%>.

Mane, P. and F. Piponnier 1997. *Dress in the Middle Ages*, trans. C. Beamish. New Haven, CT and London: Yale University Press.

The Middle English Dictionary 1998. *Middle English Compendium* online. Ann Arbor, MI: Univeristy of Michigan Press. 1 Jan. 2008. <http://quod.lib.umich.edu/m/med/>.

Newton, S. M. 1999 [1980]. *Fashion in the Age of the Black Prince: A Study of the Years 1340-1365*. Woodbridge, Suffolk: Boydell.

Norris, H. 1926. *Costume and Fashion*, Vol. 1: *The Evolution of European Dress Through the Earlier Ages*. London: Dent; New York: Dutton.

Norris, H. 1927. *Costume and Fashion*, Vol. 2: *Senlac to Bosworth 1066-1485*. London: Dent; New York: Dutton.

Norris, H. 1999 [1927]. *Medieval Costume and Fashion*. Mineola, NY: Dover.

The Oxford English Dictionary [*OED*], 2nd ed. 1989. *OED Online*. Oxford University Press. 1 Jan. <http://dictionary.oed.com/>.

Picot, E., ed., 1902-1912. *Recueil general des sotties*. Paris: Didot.

Post, P. 1910. Die französische-niederländische Männertracht einschliesslich der Ritterrüstrung im Zeitalter der Spätgotik, 1350-1475, Ein Rekonstruktionsversuch auf Grund der zeitgenössichen Darstellungen. Halle a. D. Saale, Dissertation.

Post, P. 1952. La naissance du costume masculine modern au XIVe siècle, *Actes du ler Congrès international d'histoire du costume, Venise, 31 août-7 septembre, 1952*. Venice.

Rothwell, W. 1993. From Latin to Anglo-French and Middle English: the role of the multi-lingual gloss. *Modern Language Review* 88: 581-599.

Rothwell, W. 1994. The Trilingual England of Geoffrey Chaucer. *Studies in the Age of Chaucer* 16: 45-67.

Rothwell, W. 2001. *OED, MED, AND*: The Making of a new Dictionary of English. *Anglia* 119: 527-553.

Scattergood, J. 2000. Dressing the Part in *Magnyfycence*: Allegory and Costume. *Theta* 5: *Allegory in the Theatre (L'allégorie au theatre*). Bern, etc.: Peter Lang: 55-75.

Scott, M. 1980. *The History of Dress Series, Vol. 1: Late Gothic Europe, 1400-1500*, ed. A. Ribeiro. London: Mills & Boon; Atlantic Highlands, NJ: Humanities Press.

Scott, M. 2007. *Medieval Dress & Fashion*. London: The British Library.

Smithers, G. V. 1952. *Kyng Alisaunder* [EETS]. Vol. 1: Text. London: Oxford University Press.

Stow, G. B., ed. 1977. *Historia Vitae et Regni Ricardi Secundi*. Foundation Series 21. Philadelphia: University of Pennsylvania Press.

Tait, J., ed. 1914. *Chronica Johannis de Reading et Anonymi Cantuariensis*. Manchester: Manchester University Press.

Trotter, D. 2006. Language Contact, Multilingualism, and the Evidence Problem, *The Beginnings of Standardization: Language and Culture in Fourteenth-Century England*, ed. Ursula Schaefer. Frankfurt: Peter Lang.

Twiss, T., ed. 1873. *The Black Book of the Admiralty*. Appendix – Part II. Vol. 2. Monumenta Juridica. Rerum Britannicarum Medii Ævi Scriptores. London: Longman and Trübner; Oxford: Parker; Cambridge: Macmillan; Edinburgh: A. & C. Black; Dublin: A. Thom.

Walker, G. 2000. *Medieval Drama: An Anthology*. Oxford and Malden, MA: Blackwell.

Warner, M 1983 [1981]. *Joan of Arc: the Image of Female Heroism*. Harmondsworth: Penguin.

Winser, L. 1981. *Magnyfycence* and the Characters of *Sottie*. *Sixteenth Century Journal* 12, no. 3: 85-94.

Wright, L. 1996. *Sources of London English: Medieval Thames Vocabulary*. Oxford: Clarendon.

JAVIER E. DÍAZ-VERA

Analysing the Diffusion of Scientific Metaphors through a Corpus of Middle English Medical Texts

1. Introduction

Creating metaphors is a very useful process for designating the new concepts that are constantly emerging in medicine, due to technical and scientific progress. Metaphorical conceptualization has in fact been widely recognized as a widespread phenomenon in medicine and other specialized fields, contributing to theory building and to the development of scientific thought. In the present paper, and taking the cognitive theory of metaphor (as described, among many others by Lakoff & Johnson 1980, 1999; Lakoff 1987; Johnson 1987 and Lakoff 1993), as my departure point, I am going to analyze a set of metaphors used in Middle English medical texts.

This chapter aims to describe the origins, usage and diffusion of these medical metaphors. In order to do this, I have analysed the texts contained in the Middle English Medical Texts (MEMT) corpus. This corpus, which includes 86 texts and 495,322 words from three different traditions of medical writing (surgical treatises, specialized texts, and remedy books) from 1375 to 1500, has proven to be adequate for the study of historical English, as shown by the studies carried out by Taavitsainen (2006).

My intention is twofold. On the one hand, I will identify and describe some of the metaphors recorded in Middle English texts from the different medical traditions and fields. On the other hand, I will try to show that many of the metaphors related to medicine in use in contemporary medical texts existed already, with different degrees of pervasiveness, in the Middle Ages.

This work is divided into three different parts. The first section explores some of the numerous studies on the use of metaphor in contemporary English medical texts. After that, I will describe the MEMT corpus and the type of data I am interested in for analysis. The last section offers the results and discussion of the findings.

2. Background

The study of cognitive metaphor began in non-specialized discourse and, from here, it broadened its scope to a wide variety of areas, from biology (Keller 1995) to architecture (Caballero 2006), from economy (White 1997) to technology (Cook 2007). With regard to medical language, metaphors have been considered from two different perspectives:

1. Metaphorical usages in non-specialised descriptions of the symptoms (Tompkins and Lawly 2002; Teucher 2003) and the effects (Sontag 1978, 1989; Goldbloom 2003) of illnesses and in the doctor-patient relationship (Skelton et al. 2002; Goldbloom 2003).

2. Metaphorical usages in specialised descriptions of symptoms, diseases and healing techniques (Brown 2003; Boquera 2000).

According to the studies mentioned above, whereas professionals tend to use the same set of highly standardised metaphors, independently of the degree of specialisation of the context, patients are more improvising, vivid and idiosyncratic. Consequently, doctors and patients share very few metaphors in common and, very often, this leads to difficulties in or even complete lack of mutual understanding (Sagan 1988).

Some of the highly standardised metaphors frequently used by contemporary doctors and other medical professionals have attracted the attention of many researchers from different linguistic and medical fields. This is the case of the so-called MEDICINE IS WAR metaphor, which has long informed the discourse about infectious diseases (Sontag 1978; Burnside 1983), and more recently about cancer, AIDS and other epidemic diseases (Sontag 1978, 1989; Brandt 1988; Norton et al. 1990, among many others). According to this military metaphor, which structures much of our western conception of medicine, illness treatment is conceptualised as a war led by doctors and allied health personnel.

Medical practice can also be paralleled to religious beliefs. The MEDICINE IS RELIGION metaphor probably exists from the very origins of this scientific branch and, differently to the MEDICINE IS WAR metaphor sketched above, it is shared with many other medical traditions from different parts of the world. This explains why this metaphor operates in two ways: (1) in medicine is religion (doctors have a strong vocation to serve and are full of compassion to their patients); and (2) in religion is medicine (priests heal the soul of their believers).

Another, much more recent metaphor for medicine, is gaining terrain in our conception of medicine: MEDICINE IS BUSINESS (Mustacchi and Krevans 2001). In this metaphorical view of medical practice, health professionals are seen as service providers, whereas patients are bare consumers of these services. Moreover, the various diagnostic, therapeutic, and rehabilitative resources that constitute health care represent "raw materials."

In other cases, metaphorical conceptualizations focus on a more limited aspect of medical science. This is the case of, for example, the description of pain associated to disease. Metaphoric and metonimic extensions are very common in professional descriptions of pain, as in the following examples:

i. PAIN IS A WEAPON (this is related to the general military metaphor previously described);

ii. PAIN IS A SOUND

iii. PAIN IS A QUANTITY that fills the body.

Finally, the HUMAN BODY can be referred to by health professionals in terms of such different things as a MACHINE, a LANDSCAPE (or, following the MEDICINE IS WAR metaphor, a battlefield) or a BUILDING.

3. Data and methodology

As stated above, this research is based on the MEMT corpus. In order to analyse the texts, I have used Corpus Presenter software (Hickey 2003), although manual analyses have also been performed at times. The texts included in the MEMT corpus are classified into three broad categories, according to their tradition of writing, contents and audience: surgical texts, specialised texts, and remedies and *materia medica*. This division was first suggested by Voigts (1982, 1984) and has subsequently become widely accepted.

In the first category, *surgical texts*, we have 15 texts belonging to university tradition; some of them represent the highest academic level of writing, being derived from university texts. The second category, *specialized texts*, includes 24 texts representing the academic tradition and treatises dealing with natural philosophy and reproduction, specific illnesses or fields of specialization. The third category, *remedies and materia medica*, contains texts belonging to the remedybook tradition and includes recipes, charms and herbals.

In order to extract the relevant metaphorical data from the corpus, I have either followed some of the strategies described by Stefanowitsch (2006: 2-5; given the diachronic character of our corpus, some of these strategies have been slightly adapted,) or simply created my own (Díaz Vera 2000, 2007):

- *Manual searching*: Careful reading of a selection of nine previously selected texts from the general corpus (three texts for each textual tradition).
- *Search for source domain vocabulary*: Metaphorical expressions always contain lexical items from their source

domain (for example *to fight against cancer, a pungent pain*). Rather than trying to impose present-day metaphoric patterns on medieval conceptions of medicine, this strategy will let us ascertain the existence (or not) of the metaphors described above in the Middle Ages.

- *Search for target domain vocabulary*: Here, I have searched a list of lexical items from the different medical fields and subfields in order to try to identify metaphorical usages in medical texts.
- *Search for sentences containing lexical items from both the source and the target domain* (i.e. a combination of the two methods sketched above).
- *Search for metaphors based on 'markers of metaphors'*, i.e. linguistic expressions indicating the existence of a particular metaphor, as described in the inventories proposed by Goatly (1997) Wallington et al. (2003).
- *Etymological analysis of specialized terms:* The etymological roots of scientific terms may conjure up visual images (Norri 1992:106; Díaz Vera 2007), which we will use here in order to try to reconstruct historical processes of metaphorical extension of meaning.

Through the application of these strategies, either isolated or in combination, I will propose a description of the origins, usages and diffusion of a set of metaphors used in our corpus of Middle English medical texts.

Our choice of corpus is not casual, but based on the highly specialized character of the MEMT: differently to most other historical corpora, the MEMT offers a large body of representative and monothematic texts dealing with one single target domain (i.e. medicine), allowing a systematic and highly exhaustive identification of the metaphorical expressions used in that field.

4. Middle English metaphors for medicine

4.1. The body and its parts

4.1.1. The body is a castle

Now I will analyze the results of my study of metaphorical extensions in Middle English medical texts. To start with, different metaphorical extensions used in medical descriptions of the human body indicate the existence of the metaphor THE BODY IS A BUILDING or, much more frequently, of its particular instantiation at a more specific level THE BODY IS A CASTLE.

Of special interest here is the following paragraph from *Sidrak and Bokkus*, a late-medieval work on popular belief and knowledge (including health and medicine) in question-and-answer form:

(1) *Þe herte is lord of þe body / And þe body is his wardayn / And his seruaunt wiþ al his mayn / For al þat likeþ þe herte to / It liketh to þe body also. / Þe men beth gouernours of þe herte / And þe eeren messageris smerte; / Þe tunge avokett is she / And þe sighte his sonde be; / Þe hondis beeþ hise knightes two; / Þe hertes castel is heed þerto.* (SIDRAK AND BOKKUS, P 249, f. 61v)

The metaphor THE BODY IS A CASTLE where the soul dwells has a tradition that dates as far back as Philo Judaeus (d. 40). The most often cited medieval examples of this allegory are from the writings of Anselm, St. Bernard, and Hugh of St. Victor.

St. Bernard interprets the castle as the womb of the Virgin mother, a *castellum* which Jesus entered (symbolizing the Incarnation). Later in the same sermon, St. Bernard also describes his own experience in terms of the castle metaphor, describing how his "castle" or soul had been conquered by the devil, who had taken control of all his members and destroyed the walls of Continence and Patience.

The theme was popularized by Archbishop Ralph d'Escures of Canterbury (1114-22) in a sermon which was translated into English in the early twelfth century. Thereafter it appears in numerous sermons as well as vernacular didactic literature such as Robert

Grosseteste's *Le Chasteau d'Amour*, the 13th century homily *Sawles Warde*, in *Piers Plowman* and in Chaucer's *Tale of Melibee*, among others.

The version of this metaphor found in *Sidrak and Bokkus* does not make reference to the soul at all. The heart is presented here as the lord of the castle and the body is its guardian, whereas the various bodily organs play different roles in the castle: messengers (the ears), surveyors (the eyes), advocate (the tongue) and knights (the hands). In fact the senses are not conceived of as windows used by external enemies to attack the soul, but rather as organs of the heart with a variety of functions related to mutual communication and understanding.

A similar description of the human body as a walled fortress is found, among other, in two Middle English versions of *Secretum Secretorum*: the *Secreet of Secreetes* (c1475; translated from French) and in Johannes de Caritate *The Privyte Of Privyteis* (c1484; translated from Latin):

(2) *The vertu of mannys body is fortified and perfourmed, and alle the worlde is replenysht with goodys* (SECREET OF SECREETES, f. 54v)

(3) *On wyse be deth natural þat no man may scape, þat qwan very age comyth, þe qwyche is cold and drye, þe body is destroyd* (DE CARITATE, ÞE PRIUYTE OF PRIUYTEIS, f. 53v)

(4) *And qwan þe body is depressyd, lene and weyke, or ellis drye, and þe stomake of no scharp appetyte, þan be goode to vse sotel metys þat be lyght of dygestyon* (DE CARITATE, ÞE PRIUYTE OF PRIUYTEIS f. 42v)

This architectural metaphor is exclusively used in what Taavitsainen (2006) calls 'remedies and *materia medica*', that is, recipe collections and herbals. In fact, no instances of the metaphor THE BODY IS A CASTLE have been found in the set of academic surgical texts included in our corpus.

4.1.2. The skin is a garment

This general idea of protection from the outside can also be seen in the following excerpt from a 15th century surgery treatise earlier attributed to Morstede, but now identified as a translation of Bradmore's *Philomena* (Lang 1992). The author describes the skin as a garment of woven threads of sinews and veins that cover and defend the inner body:

(5) *The skyne ys þe couerynge of þe body knyt to gedyr of threddes senews and waynys made to haue defens and fellynge ther of* (BOOK OF SURGERY, f. 11v)

The same idea, with nearly identical wording, is found in the following paragraph from the anonymous Middle English translation (end of the 15th century) of Chauliac's *Cyrurgie*, one of the most important representatives of the medieval medical learned tradition.

(6) *The skyn forsoþe is þe couerynge of þe body, iwouen of thredes of þe synowes, of þe veynes and of the arteries, imade to defendynge and 3euyng of felyng* (CHAULAC, CYRURGIE, f.11ra)

As can be seen in all these examples, the human body is conceived of by the authors of Middle English medical texts as a protective stronghold or covering that safeguards individuals from outer attacks. Quite obviously, this implies the existence of a parallel metaphorical extension in Middle English, which we will call here SICKNESS IS AN ENEMY (see 4.4. below).

4.1.3. The body is an instrument

The 17th century philosopher introduced the idea that the body was essentially a machine, while the mind possesses the controlling processes (Konopka 2002).

However, many of the academic texts included in our corpus illustrate a much earlier mechanicist vision of the human body, according to which body parts (and especially body organs with a concrete function, as opposed to the *limbs*) are conceived of as instruments or tools whose work or performance can be affected by a sickness:

(7) *Kynde makeþ instrumentis of þe body couenable to þe vertu of þe soule* (TREVISA, ON THE PROPERTIES OF THINGS 1, f. 37ra)

(8) *The ey3en beeþ the instrumentis of sight, and þai beeþ sette wiþynne þe paþþe þe whiche is a parte of þe coronal and of þe bones of þe temples* (CHAULIAC, CYRURGIE, f. 14ra)

(9) *þilke same eere-bonys // beþ holid þoru3 wiþinneforþ, þoru3 // whos hoolis passen þe senewis or // þe instrumentis of heerynge* (CHIRURGIE DE 1392, f. 19vb)

(10) *For þe werkyng of sqwyche a stomak is lyke a fyr þat on ese may brenne rede spyris and ruschys* (DE CARITATE, ÞE PRIUYTE OF PRIUYTEIS, f. 43)

(11) *It makyth febyl þe vertu natural, it makyth a man foryetful, it hurtyth alle hys fyue wyttis with þe qwyche he is gouernyd and dysposyd of alle þe bodyly werkyng* (DE CARITATE, ÞE PRIUYTE OF PRIUYTEIS, f. 62v)

Differently to the Cartesian THE BODY IS A MACHINE metaphor, the metaphoric schema used in medieval medical texts implies that each individual organ has its own functioning and is completely independent from the other organs in the body. Furthermore, following the Aristotelian tradition, the body (that is, the union of organs and limbs) is an instrument of the soul (Bos 2003: 178-179):

(12) *Þe membres and þe lymes serueþ to fele and meue and beþ þe instrumentis of þe soule, as hoond, foot, and ei3e, and oþir suche, and beþ nedeful in diuers qualitees and quantites to þe worchinge of þe soule* (TREVISA, ON THE PROPERTIES OF THINGS 1, f. 37ra).

4.2. Sicknesses and their symptoms

4.2.1. Sickness is a conqueror

As has been seen above (4.1. and 4.2), the human body is frequently referred to as protected by a protective wall (THE BODY IS A CASTLE) or by a garment (SKIN IS A GARMENT) in Middle English medical texts. This general idea of protecting someone from an enemy invasion leads us to the next metaphorical extension, according to which A SICKNESS IS A CONQUEROR.

Following this general metaphor, a sickness can come to someone's body or bodily organ (13-15) and occupies it (16-17) in order to destroy it (18-19) or, eventually be destroyed by or expelled from it (20-22).

(13) *The seconde chaptre telleth how this sikenes cometh and what is cause thereof* (JOHN OF BURGUNDY, PLAGUE TREATISE 1, p. 73)

(14) *Þan þe seknes comes thus, when þe pores are opyn for sum encheson byfore said, þan entres ayere þat es venemous* (LIBER DE DIVERSIS MEDICINIS, f. 301r)

(15) *where we schulle tretyn of seknesse þat comyth to þis membres, whyche seknesse beþ non woundes* (LANFRANC, CHIRURGIA MAGNA 2, f. 80b)

(16) *And yf it be not cut away or it ocupye the pupil and let the sy3ght, afterwarde it wyll not be easely curyd* (BENVENUTUS GRASSUS, p. 80)

(17) *werefore wan þou takest a cure be it of hy syknesse or ellis of surgery, ta keep of þe mone and of þe time wan þe seknesse took and in wat sygne it be-gan* (MEDICAL WORKS, p. 5)

(18) *The hert kyndely fleeth that is ageyne it and putteth the venym to his clensyng place and be cause that that place is stopped, that it may not oute. it passeth to the next principal parties, that is the lyver, for to destroy it* (JOHN OF BURGUNDY, PLAGUE TREATISE 1, P 73)

(19) *And if þis febylnes and indygest mater be habundaunt, þer comyth þer-of streykyng owte of þe armys and þe body, and þe bak bowyth, and þe nek, and many odyr thyngis besyde þat be contrarius to helth, destroy us of þe body, and coruptifis of natur* (DE CARITATE, ÞE PRIUYTE OF PRIUYTEIS, f. 43)

(20) *And oken apples soden in wiyn maken þe wiyn medicinable forto destroie corrupte humours and smokes þat ben in a mannes hede* (GILBERTUS ANGLICUS, COMPENDIUM, f. 55-55v)

(21) *Ete and drinke nettle seed with hony, and it wole destroye þe sikenesse þat is clepid collica passio* (MACER, DE VIRIBUS HERBARUM, f. 3a)

(22) *Hit dryith awey euyl humours of þe body, it expellith and remouith fumosite is and wyndis within the body, it makith þi mouth to be well sauourid, and all þi persone* (REGIMEN SANITATIS, f. 4)

The violence with which a sickness attacks a patient is better expressed through the metaphorical use of the verbs *to fight* and *to blast*, as in the following examples:

(23) *And þilke humours day and night / In þe body with þe good fight; / And if þe wicke maistres are, / Somtime in herte þei tourneþ 3are* (SIDRAK AND BOKKUS, f. 115v)

(24) *ffor hym þat is blastid of vnkynd eyere to do away þ=e= ach and aswage the swellynge* (LEECHBOOK 1, f. 12v)

(25) *and haue a3en her firste strenkþis of 3ongþe in þe same degree þat is in al kynde, & be mad hool parfi3tly, except þe strok of þe þundir blast, & violent brusuris, and oppressynge of to myche betynge* (QUINTE ESSENCE, f. 10)

Once again, the Middle English version of the poem *Sidrak and Bokkus* (23) illustrates perfectly this conceptual metaphoric and the general idea of imbalance between bodily humours (i.e. specific bodily fluids essential to the physiological functioning of the organism; Siraisi 1990: 103-106). Sicknesses are conceived of here as a fight between good humours and bad humours or, more specifically, as an aggression by the latter.

4.2.2. Sickness is a devourer

The second metaphor analysed here implies that some sicknesses are conceived of as wild animals that devour the human body or parts of it. This is especially true of erosive diseases, such as cancer and lupus, as illustrated by the next examples:

(26) *And generaly it availeþ in al fretyng sekene3, as in cancre, lupe, fistule, and noli-me-tangere* (ARDERNE, CLYSTERS, f. 168r)

(27) *Also þe samyn distroyeth wykked wynd cawsand ake & fret and gnawyng in þe body* (DANIEL, ROSEMARY, p. 82)

(28) *Me þinkeþ he biteth him ful sore // And freteþ and gnaweþ him wel more* (SIDRAK AND BOKKUS, f. 106r)

A nearly identical treatment is given in medieval medical texts to age, which leads us to the pervasive metaphors TIME IS A CONQUEROR and TIME IS A DEVOURER (the use of these two metaphors in literary texts is analyzed in Flores-Moreno 1999). This can be clearly seen from the following example:

(29) *For in 2 maners age cometh. In that one by due law and cours of nature, in wexyng colde, and the body wastyng and drying, and so destroying the creature* (SECRETE OF SECRETES, f. 19v)

(30) *And whan age ouercometh the body, than he most nedys dye* (SECREET OF SECREETES, f. 58v)

4.3. Medicine and medical treatments

4.3.1. Medicines are weapons

Although we have found illustrations of the SICKNESS IS A CONQUEROR metaphor in medical texts from the three textual categories described by Taavitsainen (2006), it should be mentioned here that most of the instances of this war metaphor come from academic treatises attributed to the most famous medieval physicians: John of Burgundy, Gilbertus Anglicus, Lanfranc and Chauliac.

The incidence of this metaphor in remedy books and herbals is much lower and seems to be reduced to the role of medicines and herbs in the curative process: remedies are in fact described for their usefulness against a given sickness or desease:

(31) *The vertu of þys herbe ys yf he be sode wyþ pyggys grece yt ys a gode oynment a3enst al maner of Euforbium* (AGNUS CASTUS, f. 74v)

(32) *it is a myhty medicyn ageyn alle maner seknesse in mannys body* (DANIEL, ROSEMARY, p. 82)

In academic descriptions of medicines and treatments, whenever reference to a concrete weapon or type or weapon is made, we find a strong preference for terms for defensive weapons such as *helmets*, *armours* and *shields*:

(33) *þe vtilite whi þat it was harde is be cause þat it schulde defende þe braines and þe pannikels fro diseses þat fallen oþer cummen fro wiþ-oute, & be, as it were, ane helme oþer a schelde for to defende hem* (CHAULIAC, ANATOMY (INTERPOLATED), f. 45v)

(34) *But y kytte no3t þe skyn in manere of a cros, but in þe manere of a Schelde*

oþere ellys of a thynge þat ys tryangle in þis manere (LANFRANC, CHIRURGIA MAGNA 2, f. 78b)

The adjective *defensif* is in fact frequently added to Middle English words for medicines, or even used as a noun in order to refer to a medicine or a treatment, as illustrated in the following examples from Lanfranc's *Chirurgia Magna*:

(35) *But aboute þe wounde leie a medicyn defensif, of bole armonyac, oile of rosis, & a litil vynegre; so þat (\f. 5r\) þe medicyn touche þe brynkis of þe wounde* (LANFRANC, CHIRURGIA MAGNA 1, f. 4v-5r)

(36) *be lete blood or ellis ventusid, takynge reward to alle particuler þingis tofore seid & ordeyne him a couenable dietynge, as schal be told in þe chapitle of dietynge; & leie a defensif aboute þe wounde* (LANFRANC, CHIRURGIA MAGNA 1, f. 27v)

(37) *and aboue þe wounde a defensiff of bol armonyac tofore seyd, tyl þat þe quyter be engendryd* (LANFRANC, CHIRURGIA MAGNA 2, f. 78v)

5. Conclusions

The medical metaphors found in our Middle English corpus draw upon a variety of experiential domains, among which the most recurrent are those of *war*, *architecture*, *zoology*, *clothing* and *machinery*.

The semantic extensions described here illustrate what Lakoff and Turner (1989) and Lakoff (1993) call *conceptual metaphors,* that is, mapping a concept onto another one. These can be further classified into:

1. *Ontological* metaphors, whereby events, activities, emotions, ideas, etc. are viewed as entities and substances (e.g. MEDICINES ARE WEAPONS); and

2. *Structural* metaphors, whereby a concept is metaphorically structured in terms of another (e.g. GETTING SICK IS BEING CONQUERED).

Ontological schemas frequently depend on a general, structural schema subsuming them. This is the case of, for example, MEDICINES ARE WEAPONS, SICKNESS IS A CONQUEROR or THE BODY IS A CASTLE, all of which belong to the more general schema MEDICINE IS WAR. These can be decomposed into different parts or metaphors that focus on various aspects of medical activity (for instance the body, the patient, the treatment) and, as has been seen in many of the examples above, frequently co-occur in the same paragraph.

Furthermore, through the analysis of the metaphorical extensions found in our corpus of Middle English texts, I have been able to show that many of the contemporary metaphors described by the beginning of this paper existed already in the Middle English period. This is especially true of two of the dominant medical metaphors: THE BODY IS A BUILDING and MEDICINE IS A WAR.

However, as has been noted above, the medieval versions of these two metaphors emphasize the defensive aspects of the body and of medicine in general, in contrast with the more aggressive role of sicknesses and diseases. Rather than fighting against disease, one could say that medieval doctors treated their patients by trying to strengthen their physical defences (i.e. their body) or to expel the enemy (i.e. the sickness) once the attack had started.

References

Boquera Matarredona, M. 2000. *La Traducción de Metáforas en un Texto de Divulgación Médica*, Ibérica 2, 13-25.

Bos, Abraham P. 2003. *The Soul and Its Instrumental Body: A Reinterpretation of Aristotle's Philosophy of Living Nature.* Leiden/Boston: Brill Academic Publishers.

Brandt, Allan M. 1988. AIDS and Metaphor: Toward the Social Meaning of Epidemic Disease. *Social Research* 45, 413-432.

Brown, T. 2003. *Making Truth: Metaphor in Science.* Urbana: University of Illinois Press.

Burnside, John W. 1983. Medicine and War: A Metaphor. *Journal of the American Medical Association* 15, 2091.

Caballero, Rosario 2006. *Re-Viewing Space: Figurative Language in Architects' Assessment of Built Space*. Berlin/New York: Mouton de Gruyter.

Childress, James F. 1989. Problems, Principles and Metaphors in Biomedical Ethics. In Mitchell, Carlton T. (ed.) *Values in Teaching and Professional Ethics*. Macon, GA: Mercer University Press.

Cook, Jennifer C. 2006. *Machine and Metaphor: The Ethics of Language in American Realism.* London: Routledge.

Díaz-Vera, Javier E. 2000. "*On sarnese pu acenst cild*: Being Born in the History of English, with Special Reference to the Old and Middle English Periods. *Estudios Ingleses de la Universidad Complutense*, 8, 79-96.

——— 2007. Metaphors We Learnt By: Cultural Traditions and Metaphorical Patterns in the Old English Vocabulary of Knowledge. *Revista Canaria de Estudios Ingleses,* 55, 99-106.

Flores-Moreno, Cristina. 1999. Life, Tima and Death in Shakespeare's *Sonnets*: The Lakoffian Approach to Poetic Metaphors, *RESLA* 7, 287-304.

Goatly, Andrew 1997. *The Language of Metaphors*. London: Routledge.

Goldbloom, D. S. 2003. Language and metaphor. *Canadian Psychiatric Association Bulletin*. Toronto, Ontario.

Harley, David N. 1993: Medical Metaphors in English Moral Theology. *Journal of the History of Medicine and Allied Sciences*, 48(4), 396-435.

Hickey, Raymond 2003. *Corpus Presenter*. Amsterdam and New York: John Benjamins.

Johnson, Mark 1987. *The Body in the Mind.* Chicago: The University of Chicago Press.

Keller, E.F. 1995. *Refiguring Life: Metaphors in Twentieth Century Biology*. New York: Columbia University Press.

Konopka, Andrzej K. 2002. Grand Metaphors of Biology in the Genome Era. *Computers & Chemistry* 26, 397-401.

Lakoff, George 1987. *Women, Fire, and Dangerous Things: What Categories Reveal About the Mind*. Chicago: The University of Chicago Press.

Lakoff, George 1993. The Contemporary Theory of Metaphor. In Ortony, A. (ed.) *Metaphor and Thought*. Cambridge: Cambridge University Press.

Lakoff, George / Johnson, Mark 1983. *Metaphors We Live by*. New York: Basic Books.

——— 1999. *Philosophy in the Flesh: The Embodied Mind and Its Challenge to Western Thought*. New York: Basic Books.

Lakoff, George / Turner, M. 1989. *More than Cool Reasons: A Field Guide to Poetic Metaphor*. Chicago: The University of Chicago Press.

Lang, S.J. 1002. John Bradmore and His Book *Philomena*. *Social History of Medicine* 5, 121-130.

Mustacchi, Piero / Krevans, Julius R.. 2001. Medicine and Money: Metaphorically Speaking the Metaphor of Health Care Provision as a Factory. *Western Journal of Medicine*, 175, 14–16.

Norri, Juhani 1992. *Names of Sicknesses in English, 1400-1550: An Exploration of the Lexical Field.* Helsinki: Suomalainen Tiedeakatemia.

Norton, Robert / Schwartzbaum, Judith / Wheat, John 1990. *Language Discrimination of General Physicians: FIS Metaphors Used in the AIDS Crisis,* Communication Research 17.6, 809-826.

Sagan, Leonard A. 1998. *The Health of Nations: True Causes of Sickness and Well-Being*. New York: Basic Books.

Siraisi, Nancy G. 1990. *Medieval and Early Renaissance Medicine: An Introduction to Knowledge and Practice*. Chicago: The University of Chicago Press.

Skelton John R. / Wearn, Andy M. / Hobbs, F. D. Richard 2002. A Concordance-based Study of Metaphoric Expressions Used by

General Practitioners and Patients in Consultation, *The British Journal of General Practice* 52(475), 114-118.

Sontag, Susan 1978. *Illness as Metaphor*. New York: Farrar, Straus and Giroux.

Sontag, Susan 1989. *AIDS and its Metaphors*. New York: Farrar, Straus and Giroux.

Stefanowitsch, Anatol 2006. Corpus-based Approaches to Metaphor and Metonymy. In Stefanowitsch, Anatol and Gries, Stefan Th. (eds) *Corpus-based Approaches to Metaphor and Metonymy*. Berlin and New York: Mouton de Gruyter.

Teucher, Ulrich 2003. *Metaphor in Crisis: The Language of Suffering*. University of British Columbia. Available at <http://www.english.ubc.ca/projects/pain/dteuch.htm>.

Taavitsainen, Irma 2006. Thought-styles and Metadiscourse in Six Hundred Years of Medical Writing. In Gotti, Maurizio / Salager-Meyer, Francoise (eds) *Medical Writing*. Bern: Peter Lang.

——— / Pahta, Päivi / Mäkinen, Martti 2005. *Middle English Medical Texts*. Amsterdam: John Benjamins.

Tompkins, Penny / Lawley, James 2002. The Mind, Metaphor and Health. *Positive Health,* 78. Available at <http://www.positivehealth.com/article-abstract.php?articleid= 655>.

Voigts, Linda Ersham 1982. Editing Middle English Medical Texts: Needs and Issues. In Levere, Trevor (ed.) *Editing Texts in the History of Science and Medicine.* New York: Garland.

——— 1984. Medical Prose. In Edwards, A. (ed.) *Middle English Medical Prose: A Critical Guide to Major Authors and Genres*. New Jersey: Rutgers University Press.

Wallington, A.M., Barnden, J.A., Barnden, M.A., Ferguson, F.J. & Glasbey, S.R. 2003. *Metaphoricity Signals: A Corpus-based Investigation.* Technical Report CSRP-03-05, School of Computer Science, The University of Birmingham, U.K.

White, Michale 1997. The Use of Metaphor in Reporting Financial Market Transactions. *Cuadernos de Filología Inglesa*, 6.2, 233-245.

Winslow, Gerald R. 1996. Minding our Language: Metaphors and Biomedical Ethics. In Shelp E.E. (ed.) *Secular Bioethics in Theological Perspective*. Boston: Kulwer Academic.

María Laura Esteban-Segura

Punctuation practice in G.U.L. MS Hunter 509[1]

1. Introduction

Punctuation is, among the different elements which conform the study of palaeography, one of the most rudimentary, although sometimes it is also one of the most confusing (Petti 1977: 25). Its standardization, in a similar line to that of the spelling system, starts with the printing press, so that in manuscripts punctuation is employed in a less consistent way than in more modern books. A quick look at medieval manuscripts may lead to think that marks are used arbitrarily. However, although punctuation is not always systematic, as single manuscripts may present their own inventory of symbols, different uses and functions can be established, which may help in the elucidation of the text.

As for studies about punctuation systems in Middle English, they have been traditionally characterized by both their scarcity and the lack of scholarly agreement. Gradon (1983: 39) remarks that "the punctuation of medieval manuscripts has long been a matter of contention, speculation or even despair". However, there has been a growing interest in the area started by the contributions of Lucas (1971) and Arakelian (1975), which has continued over recent years with the works of scholars who have also taken an interest in scientific and technical texts on the vernacular (Rodríguez-Álvarez 1999; Alonso-Almeida 2001, 2002-2003; Calle-Martín 2004; Calle-Martín

1 The present research has been funded by the Spanish Ministry of Science and Education (grant number HUM2004–01075FILO) and by the Autonomous Government of Andalusia (grant number HUM-2609). These grants are hereby gratefully acknowledged.

and Miranda-García 2005; Obegi-Gallardo 2006; Marqués-Aguado 2008: 39-47).

According to Parkes (1978: 132-133), there are several factors governing the use of punctuation in the different periods of the Middle Ages. They include the nature of the text, the manner in which this was read and understood by scribes, or its purpose. These issues are important and related to the question of modernizing punctuation in editions of medieval texts, which is a controversial topic. Should the editor reproduce the original punctuation or provide a new modern system? Heyworth (1981: 141) argues that the lack of awareness of punctuation marks and the modern treatment of them may explain the standards in editorial punctuation "which would be tolerated in no other branch of medieval text-critical studies".

The main objectives of this article are to offer an account of the punctuation practice in a specific medical text, G.U.L MS Hunter 509, and to establish, if possible, the most appropriate equivalent of these marks in modern punctuation. In the next sections, the text under investigation will be briefly described and the methodology explained. Then the inventory of marks of punctuation will be analysed and the most important uses discussed. Finally, the conclusions will be provided.

2. Description and methodology

The manuscript under consideration, Hunter 509, is housed at the Hunterian Collection in the Special Collections Department of Glasgow University Library. I have worked on its edition and study for my PhD thesis (Esteban-Segura 2008) and it also forms part of a research project carried out by the research group HUM-497 from the University of Málaga, in collaboration with the Universities of Oviedo and Glasgow. The project involves the electronic editions of the Late Middle English scientific texts belonging to the Hunterian Collection.[2]

---

2 <http://hunter.filosofia.uma.es/manuscripts>.

The folios selected for the analysis are 14r-167v. They are based on the Middle English Gilbertus Anglicus, an adaptation written in prose around 1400 of the *Compendium medicinae*, which was compiled by Gilbertus Anglicus in Latin around 1240. It was one of the standard references for physicians until the seventeenth century and also an academic text. Hunter 509 was copied during the second half of the fifteenth century by Thomas Westhaugh, a Doctor of Divinity who graduated from Cambridge University and was later Confessor General of Syon Monastery (Esteban-Segura 2008: 57).

As for the methodology, the first stage involved the transcription of the text. Part of this was done directly from the manuscript, but the microfilm was also employed for the process. The digitized images were available later on, and were used for revision. Once the text was transcribed, instances of each mark were classified and grouped manually in order to describe its main uses.

3. Analysis

3.1. Inventory of punctuation marks

The inventory of symbols in Hunter 509 includes the *punctus*, the *punctus elevatus*, the virgule, the colon, the paragraph mark, and the caret. From f. 120v onwards, punctuation marks become scarce as they are less frequently employed. A change of author in the source text from which the scribe was copying Hunter 509 could be an explanation for this change. However, a lexical study would be needed in order to confirm this hypothesis.

A very general distinction has to be made between grammatical and rhetorical punctuation. The first type provides syntactic sense and is used to mark structural relations among sentence constituents. Rhetorical punctuation, on the other hand, helps to signal necessary pauses in an oral exposition. It is common that both types of punctuation coincide, since the pauses made when speaking normally take place at the end of a syntactic unit. The fact that the manuscript under study was a reference work for medical practice indicates a grammatical function of punctuation.

3.2. The punctus

The most employed mark is the *punctus*, which appears in raised position (·). It has a number of different uses, which include signalling sentential, clausal and phrasal relationships.

The *punctus* at sentence level is used to associate or to split structurally independent units. These can be divided into a set of sub-functions, such as to mark off the end of a sentence. In the examples provided hereafter, the original mark appears first and then the modern equivalent suggested (both in bold). A full stop would be the most appropriate counterpart in the following case (example 1):

(1) "as it was bie fore seid in / in oþ*ere*· greuaunces of þe breste **·** a · Mannes/ herte/ þ*at* is þe pr*in*cipal · membre of · a · mannes body" (ff. 88r-88v)
"as it was bie-fore seid in oþ*ere* greuaunces of þe breste**.** A mannes herte, þ*at* is þe pr*in*cipal membre of a mannes body"

The *punctus* is also employed to introduce coordinate clauses. It signals paratactic clauses introduced by *and* or by correlative elements. The modern rendering would be a comma (example 2) if the sentences are semantically linked or a semicolon or full stop (example 3) if they are not.

(2) "oþ*er*while it is be fore **·** oþ*er*whiles it is be hynde" (f. 123r)
"oþ*er*while it is be-fore**,** oþ*er*whiles it is be-hynde"

(3) "And let purgen hym w*ith* · a · calogoge **·** And lat hi*m* vsen · an electuarie" (f. 68r)
"and let purgen hym w*ith* a calogoge**.** And lat hi*m* vsen an electuarie"

Another use is to connect the main and the subordinate clause in the following types of subordination: object (example 4), adjectival (example 5) and adverbial clauses, such as those of condition (example 6), cause, purpose or time.

(4) “And vnd*er*stond · þ*at* alle þe medicines þ*at* schulli*n* bie” (f. 46r)
“And vnd*er*stond þ*at* alle þe medicines þ*at* schulli*n* bie”

(5) “And siþþe vse colliries · þ*at* bien · a · lytil · corrosyf” (f. 39r)
“and siþþe vse colliries þ*at* bien a lytil corrosyf”

(6) “And · ȝif þe ache bie ful greuous · tak oile of roses” (f. 47v)
“And ȝif þe ache bie ful greuous, tak oile of roses”

The modern rendering will depend on the type of clause and subordinator; in conditional clauses, for instance, the comma would be the most appropriate counterpart.

Another use of the *punctus* at sentence level is to call attention to what follows (example 7). A suitable equivalent would be the colon.

(7) “and it is þus j maad · tac/ of Mirabolan*e* of kebuli indj” (f. 134v)
“and it is þus j-maad: Tac of mirabolan*e*, of kebuli indj”

The *punctus* employed at clause level serves several purposes. One of them is to separate clause constituents, such as the subject and the verb (example 8).

(8) “Good siripis · bien þe sirip · of i*n*nibes oþ*er* of rosees” (f. 78r)
“Good siripis bien þe sirip of i*n*nibes, oþ*er* of rosees”

Another one is to signal the coordination of phrases (example 9).

(9) “*and* to smyte · drie//nesse of þe mouþ blacnesse of þe tu*n*ge · mych*e* scharp · *and* bytt*er* ache · *and* mych*e* stery*n*ge of þe hert” (f. 17vbis)
“*and* to smyte, drienesse of þe mouþ, blacnesse of þe tu*n*ge, mych*e* scharp *and* bytt*er* ache, *and* mych*e* stery*n*ge of þe hert”

The *punctus* is also used to enumerate:

(10) “cardamome · origane · anet carewey fenel · silere mou*n*tayn · comyn · sauge” (f. 43r)
“cardamome, origane, anet, carewey, fenel, silere mou*n*tayn, comyn, sauge”

In the last two examples, if modernized, the comma would be the most suitable counterpart.

The *punctus* is used at phrase level to circumscribe words, such as determiners (figure 1), apothecaries' weights (figure 2), the ampersand (figure 3), or prefixes (figures 4, 5), as well as numerals.[3] The *punctus* acts here as a linking device and also as a way of avoiding confusion. As for a modern equivalent, no alternative mark is needed, except in the case of prefixes where a hyphen could be employed.

Figure 1

Figure 2

Figure 3

Figure 4

Figure 5

3.3. The punctus elevatus

The *punctus elevatus* (:') is less employed than the *punctus* in the text. It indicates a lighter pause than a full stop, and ranges in use from the current comma to the semicolon. One of its uses at sentence level is to introduce coordinate clauses (example 11).

(11) "ffor þe medicine is but an help :' and þe p*ri*ncipal · is þe me*m*bre þ*a*t is siyk //" (f. 31r)
"for þe medicine is but an help, and þe p*ri*ncipal is þe me*m*bre þ*a*t is siyk."

3 Special thanks must go to David Weston, Keeper of the Special Collections Department of Glasgow University Library, for kindly allowing the reproduction of images.

Besides, it can separate subordinate clauses, especially adverbial clauses, and among these, the most frequent ones are those of condition (example 12) and of time (example 13). The equivalent would be a comma on most occasions.

(12) "And ȝif it be of male*n*colie ⸮ þ*o*u maist/ knowe*n* it bi blacnesse · *and* leeneschipe" (f. 23r)
"And ȝif it be of male*n*colie, þ*o*u maist knowe*n* it bi blacnesse, *and* leeneschipe"

(13) "ffor wha*n*ne · a ma*n* leiþ · his hand on þe herte ⸮ he trembleþ" (f. 91v)
"for wha*n*ne a ma*n* leiþ his hand on þe herte, he trembleþ"

Or else, it can introduce an explanation (example 14), in which case a comma would be the best rendering.

(14) "þ*at* is to sey ⸮ to smyty*n* · a · ȝey*n* fro þe eiȝe" (f. 30v)
"þ*at* is to sey, to smyty*n* a-ȝey*n* fro þe eiȝe"

This mark may also be used to call attention to what follows (example 15); the colon could be employed as a modern equivalent.

(15) "And w*ith* þis oynemen*t* ⸮ tac of euforbe · piretr*e* · seneuey" (f. 21r)
"and w*ith* þis oynemen*t*: Tac of euforbe, piretr*e*, seneuey"

The *punctus elevatus* at clause level is used to join correlative elements (example 16).

(16) "And þ*er*for*e* it nedeþ to haue · co*n*fortatyues in þe be gy*n*//nyng/· þ*at* bien neþ*er* ful hoote ⸮ naþ*er* ful coolde" (f. 31r)
"And þ*er*for*e* it nedeþ to haue co*n*fortatyues in þe be-gy*n*nyng þ*at* bien neþ*er* ful hoote naþ*er* ful coolde"

It can also separate subject and verb (example 17), and phrases within the clause (example 18).

(17) "And sich*e* stufes oþ*er* fumigacyou*n*s ⸮ xulle*n* bie of swyche þi*n*ges" (f. 46r)
"And sich*e* stufes oþ*er* fumigacyou*n*s xulle*n* bie of swyche þi*n*ges"

(18) "þe ache ∵ on þe on syde · ~~onf~~ of þe heed" (f. 16r)
"þe ache on þe on syde of þe heed"

In general, no punctuation mark is needed when modernizing in the last two instances.

3.4. The virgule

The virgule or oblique stroke (/) is very frequent in the manuscript. Its use is similar to that of the comma (example 19).

(19) "*and* þ*er*e ben · in his vrine smale · stremys / *and* ȝif þe colo*ur* of þe vrine be blac" (f. 12v)
"*and* þ*er*e ben in his vrine smale stremys, *and* ȝif þe colo*ur* of þe vrine be blac"

It is also employed to mark the end of words (example 20).

(20) "goode drynckys/ *and* · leuyng/" (f. 101v)
"goode drynckys *and* leuyng"

Different degrees of emphasis are accomplished by doubling it or combining it with the *punctus*. The virgule combined with the *punctus* usually denotes the end of a major unit of sense and, therefore, a full stop would be the most appropriate counterpart in modern punctuation. In these cases, the virgule can appear followed by a *punctus* and another virgule (example 21), and also by a *punctus* (example 22).

(21) "let hym vsen coold siripis/ as of violet/ *and* wat*er* also þ*a*t þe sedis of coold herbis · bien j soden i*n*ne /·/ Also seeþ · þe iuys of plau*n*teyn · tyl þe halfendel · bie j sode*n* · a · wey ·" (f. 109r)

(22) "þese vrines teche to know · al man*er* sicnesse · of þe mawe /· Wha*n* þe vrine is riȝt þi*n*ne · *and* liyk · pale" (f. 13r)

Or it can appear preceded by a *punctus* (example 23).

(23) "Oþ*er* þu myȝtest make · þi suppositorys · of boles galle · sal ge*mm*e · *and* mouse drit/ j · medlyd to gedere ·/ But suppositories · for · childre*n* xul be maad w*ith* lard" (f. 22r)

The double virgule (//) is used at the end of the line to indicate that the word does not end there and that the element in the next line belongs to it (example 24).

(24) "arn wele j soden · cast þ*er* to þe · v · part of vi*n* //
egre · *and* · lat hym sittyn · a · good while in þ*at*" (f. 117r)

This can also be represented (although not so commonly) by the *punctus* followed by a double virgule (example 25) or by the *punctus* alone (example 26).

(25) "hen · oþ*er* · of · a · chike · And late hi*m* vsen · peny ·//
des *and* · amyde · *and* · applen j · rosted · *and* · bot*e*re w*ith*" (f. 69v)

(26) "*and* bie laxe · *and* bryng//<e>þ here in to · a · palesie · *and* ma ·
keþ · a · ma*n* to lese hese speche · Oþ*er*while it falleþ" (f. 52v)

Sometimes no punctuation mark is employed to signal this (example 27).

(27) "haue no feu*ere* · lat here haue · a · baaþ of st*ri*cto
ries · *and* ellys late wetyn · a · clooþ in water*e* þat" (f. 93v)

In several instances, on the other hand, the double virgule is employed at the end of the line even if the word does not continue in the following one (example 28).

(28) "And wha*n* it keleþ · cast i*n*ne þ*in* poudres · *and* medle //
it al to gidere · // Anoþ*er* hoot dragagant tak of" (f. 70v)

It can also appear at the beginning of the line to mark a grammatical pause (example 29).

(29) "til · *and* liþ dou*n* riȝt/ · þese been toknes/ of deeþ · // And ȝif he haþ not/ þese signes ∴ ȝif hi*m* pilules ·" (f. 20r)

The double virgule can be preceded by the *punctus*, in which case it signals the end of a major unit of sense as when, for instance, a recipe finishes (example 30).

(30) "deliȝ · *and* · oile · of violet/ · oþ*er* botere j medlid w*ith* whete mele ·// And ȝif al þis ne helpeþ nat *and*" (f. 67r)

The same type of pause can be indicated by means of the double virgule without the *punctus* (example 31).

(31) "*and* þe þrote · þ*a*t bie*n* þe instrume*ntes* of þe vois // And · ȝif it comeþ of blood · þese bien þe tokenes" (f. 67v)

3.5. Other punctuation marks

The colon, whose use is not very frequent, serves as an intermediate pause (example 32), representing sometimes an alternative to the *punctus* as, for instance, with numerals (example 33):

(32) "And wheþ*er* it bie of blood oþ*er* of colre : let p*ur*ge*n* hi*m* wiþ su*m* galogoge" (f. 71v)

(33) "lat hem ligge in reste : xl · daies" (f. 159v)

Paragraph marks, employed to introduce a new paragraph, are confined to a few folios (60v-62r and 148v-149r). They are in red ink (figure 6) and so is the underlining of headings in the same folios (figure 7). They were probably added later as the original punctuation can be distinguished underneath. It is difficult to establish why this section alone was rubricated since, as Jones (2000: 210) notes, it crosses a quire boundary, does not start or end at a quire boundary, and also begins and ends in the middle of a sentence.

Figure 6

Figure 7

The caret, which occurs as two oblique strokes, indicates interlineation. It is usually inserted below the line of writing to signal the place where a missing letter or word should go (figure 8); occasionally, it is forgotten (figure 9).

Figure 8

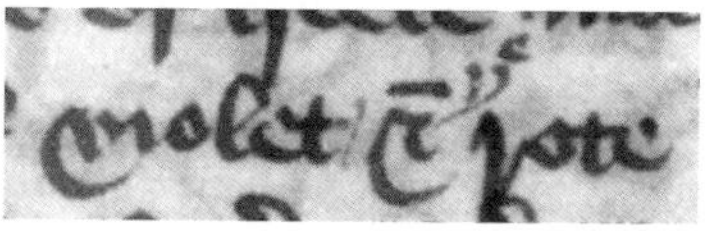

Figure 9

3.6. Other uses

Apart from the main use of the *punctus* in punctuation, some vowels may appear dotted, especially if they carry an expansion mark (figures 10-12).

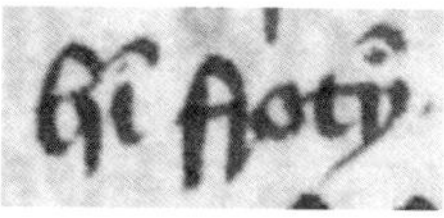

Figure 10

Figure 11

Figure 12

In the cases of *y* and *u*, this can help to avoid confusing them with *þ* and *n* respectively, as the graphs for each pair are identical.

On other occasions, when capital letters do not occur at the beginning of a section or of a line but in the middle of it, a dot may be included within the letter (figures 13-14).

Figure 13

Figure 14

The triangle of dots or triple period (sometimes called *periodus*) is frequently used in the annotations done in the margins and it is also found at the end of the manuscript to mark the ending of the treatise (figure 15). This works at the macro-structural level.

Figure 15

A diacritic sign (´) is placed over the letter *i* to avoid confusion with other minims (figure 16).

Figure 16

As for the use of capital letters, they usually appear after a major pause, although they can be found in the middle of the sentence. There are only a few rubricated initials (figure 17).

Figure 17

4. Conclusions

One of the main problems posed by punctuation in Hunter 509 is to determine if the copying of the text and the punctuation marks are contemporary, that is, if the scribe was punctuating at the time of writing, or if they were added later. Sometimes added or modified marks of punctuation are easy to identify because they appear in darker ink. From the evidence supplied by the manuscript, it seems that punctuation was done alongside the text. Another difficulty found in the manuscript is to distinguish actual marks from ink splashes or stains of the parchment on which the text was written.

Punctuation marks are helpful in order to understand a text, as they show the relationship between phrases, clauses, and sentences. Punctuation also contributes to a better reading of the text. In this study, the uses of the different marks have been classified to find out if they follow a defined pattern. From the analysis, it may be drawn that the scribe generally seems to follow a more or less consistent punctuating pattern, even though there are cases where uses overlap (for instance, coordination is introduced by the *punctus* and the *punctus elevatus*). Their functions are mainly grammatical.

As for the question of modernizing, this has been treated independently with some uses and maintaining the original punctuation or changing it will depend on the purpose or scope of the edition.

References

Alonso-Almeida, Francisco 2001. Punctuation Practice in a Late Medieval English Medical Remedybook. *Folia Linguistica Historica* 22/1-2, 207-232.

—— 2002-2003. An Assessment of Punctuation in Some Middle English Gilbertus Anglicus's Versions of 'the Sekenesse of Wymmen'. *Philologica canariensia* 8/9, 13-42.

Arakelian, Paul G. 1975. Punctuation in a Late Middle English Manuscript. *Neuphilologische Mitteilungen* 76, 614-624.

Calle-Martín, Javier 2004. Punctuation Practice in a 15th-Century Arithmetical Treatise (MS Bodley 790). *Neuphilologische Mitteilungen* 4, 407-422.

Calle-Martín, Javier / Miranda-García, Antonio 2005. Editing Middle English Punctuation. The Case of MS Egerton 2622 (ff. 136-165). *International Journal of English Studies* 5, 27-44.

Esteban Segura, M. Laura 2008. *G.U.L. MS Hunter 509 (ff. 1r-167v): An Edition and Philological Study*. Unpublished PhD thesis. Málaga: Universidad de Málaga.

Gradon, Pamela 1983. Punctuation in a Middle-English Sermon. In Stanley, Eric Gerald / Gray, Douglas (eds) *Five Hundred Years of Words and Sounds*. Cambridge: A. S. Brewer, 39-48.

Heyworth, Peter L. 1981. The Punctuation of Middle English Texts. In Heyworth, Peter L. (ed.) *Medieval Studies for J. A. W. Bennett*. Oxford: Clarendon Press, 139-157.

Jones, M. Claire 2000. *Vernacular Literacy in Late-Medieval England: The Example of East Anglian Medical Manuscripts*. Unpublished PhD thesis. Glasgow: Glasgow University Library.

Lucas, Peter J. 1971. Sense-Units and the Use of Punctuation-Markers in John Capgrave's Chronicle. *Archivium Linguisticum* 2, 1-24.

Marqués-Aguado, Teresa 2008. *Edition and Philological Study of G.U.L. MS Hunter 513 (ff. 37v-96v)*. Unpublished PhD thesis. Málaga: Universidad de Málaga.

Obegi-Gallardo, Nadia 2006. Punctuation in a Fifteenth-Century Scientific Treatise (MS Cambridge L1.4.14). *Linguistica e Filologia* 22, 99-114.

Parkes, Malcolm B. 1978. Punctuation, or Pause and Effect. In: Murphy, James J. (ed.), *Medieval Eloquence. Studies in the Theory and Practice of Medieval Rhetoric*. Berkeley, Los Angeles: University of California Press, 127-142.

Petti, Anthony G. 1977. *English Literary Hands from Chaucer to Dryden*. London: Edward Arnold.

Rodríguez-Álvarez, Alicia 1999. The Role of Punctuation in 15th-Century Vernacular Deeds. *Folia Linguistica Historica* 19/1-2, 27-51.

TERESA MARQUÉS-AGUADO

The Dialectal Provenance of G.U.L. MS Hunter 513 (ff.1r – 37r)[1]

1. Introduction

Middle English displays a great deal of written variation, and as a matter of fact, it has been characterized as "a time of unsettled orthography" (Milroy 1992: 200). In an attempt to cater for this variation, Moore, Meech and Whitehall (1935) initially devised a method to study the dialectal provenance of Middle English texts, although their approach showed several shortcomings, such as the limited use of literary texts, the non-representative nature of some of the texts, the wide timespan they aimed at covering, and their disregard for translated texts (Fernández 1982: 603; McIntosh 1989a: 24–28).

These shortcomings were taken as the starting point by the authors of the *Linguistic Atlas of Late Mediaeval English* (*LALME*), who particularly disagreed with the widespread idea of the time that the written language was a debased reflection of the spoken language. Therefore, one of the methodological primes of *LALME* was that the written and the spoken language had to remain apart, and that the former could be studied independently (McIntosh 1989b: 11).

In this line, Benskin proposed that "regardless of whether it corresponds to variation in the spoken language, orthographic

1 The present research has been funded by the Spanish Ministry of Science and Innovation (FFI2008-02336/FILO, "Grupo Consolidado"), by the autonomous government of Andalusia (Proyecto de Excelencia P07-HUM–02609), and by the Ministry of Education (FPU programme, reference number AP–2004–0244). These grants are hereby gratefully acknowledged.

variation has its own regional patterns" (1991: 24), which are the ones under study in *LALME*.

The other basic tenet underpinning the methodology of *LALME* is the value of translated texts, which had been previously considered to be of little or no value for dialectal studies. Accordingly, "the [translated] text is not a debased form of the dialect of the original, but a valid representation of the dialect of the copying scribe" (Laing 1989: x). This is the case of the text under study, which is a Middle English translation of a Latin ophthalmological treatise preserved in Glasgow University Library, MS Hunter 513 (ff. 1r – 37r).

It must be added, though, that most of the analyses carried out this far in terms of dialectal provenance have focused on literary texts (for instance, Gower [Samuels and Smith 1981] or Chaucer [Samuels 1983]), although growing attention is being recently paid to texts belonging to the field of *Fachprosa*.[2] The present study, therefore, follows Dossena and Lass' suggestion that texts other than literary should be examined from a dialectal perspective (2004: 11).

In the light of this, our aim is to present an analysis of the written variants found in the medical treatise under study in an attempt to provide a likely place of provenance.

For the purpose, this article is structured into four sections: first, some information as regards the text and its relationship to the other extant witnesses; second, an explanation of the methodology followed; third, the analysis of the dialectal provenance of the ophthalmological treatise; and finally, the conclusions drawn from such study.

---

2 See for instance the analyses of a treatise on mensuration by Calle Martín (2000), or of several witnesses of Guy de Chauliac's *Chirurgia Magna* by Taavitsainen (2004), among others. More recently, Esteban Segura has carried out a dialectal analysis of a Middle English version of Gilbertus Anglicus' treatise (2008: 375-378).

2. The text

The text under analysis is a mid-fifteenth century copy held in one of the manuscripts of the Hunterian Collection in Glasgow University Library (MS Hunter 513, ff.1r – 37r). This codex is a medical miscellany, as it contains other three medical texts, that is, an antidotary and two pseudo-Hippocratic texts (Young and Aitken 1908: 421).

As mentioned above, the text under consideration is one of the Middle English translations of the treatise known as *De Probatissima Arte Oculorum*, which was probably written by a Benvenutus Grassus. Grassus is speculated to have lectured at Montpellier, to have travelled far and wide across Europe and Northern Africa, and to have belonged to the Salernitan school, although none of these conjectures can be confirmed (Eldredge 1996: 1–5; Eldredge 2002: 155–167; Marqués Aguado *et al.* 2008: 9–14). In this line, the only piece of evidence that has come down to us is this treatise, and, as a matter of fact, Benvenutus' impact on other contemporary medical writers can be questioned, as he is not cited very often in other treatises of the time.[3]

When coming to the vernacular version, no clear Latin exemplar has been identified (Eldredge 1996: 36). Nonetheless, four Middle English witnesses are found: Glasgow, University Library, MS Hunter 503 (*H1*); Glasgow, University Library, MS Hunter 513 (*H2*);[4] Oxford, Bodleian Library, MS Ashmole 1468 (*A*); and London, British Library, MS Sloane 661 (*S*). According to Eldredge (1996: 20–37), these are organized into two branches: on the one hand, *H2* and *A* (the latter being probably the exemplar from which the former was copied, despite being fragmentary); and on the other, *H1* and *S* (both of which seem to derive from the same exemplar, the latter being also fragmentary). Eldredge himself has done some research on the

3 Some exceptions are Jean Yperman (Tabanelli 1969: 144, 162–163), Chauliac (Ogden 1971: 634) and the *Antidotary* that follows this *Treatise on the Eye*, in ff.37v – 96v (Marqués Aguado 2008: 161). See also Eldredge (1998: 48) for a discussion of the nature of the *Treatise on the Eye*.

4 An edition of this witness has been recently published (Marqués Aguado *et al.* 2008).

dialectal provenance of the four witnesses, but the area he suggests is not very precise, as *H2* is considered to be representative of the dialect of the Midland area (1996: 44).

3. Methodology

In order to determine the geographical location of this treatise, the well-known 'fit'-technique (McIntosh *et al.* 1986, 1: 10–12) has been followed, whereby the likely provenance of a text may be established in view of the evidence rendered by the group of anchor texts, with which it is compared. The procedure is as follows (McIntosh *et al.* 1986; Benskin 1991): first, the set of items to be included in the questionnaire (or 'profile') is selected; second, these are plotted in the dot maps, thus providing the approximate geographical area where the text was produced; third, the results rendered are compared with the item maps, which show in greater detail the areas where particular variants are recorded; finally, and if necessary, recalcitrant and other unattested forms can be checked against the county dictionary (Benskin 1991: 24). Nonetheless, it must be born in mind that unparalleled forms may appear, coming from either neighbouring dialectal areas or from other dialects from further apart. Simple as it may look, one of the main problems is determining the items that should come in the questionnaire, as these must be representative enough to trigger dialectal information. These must be selected on the grounds of two criteria: their frequency and the degree to which they display dialectal variation.

In the present case, the task of retrieving the items for the profile has been substantially eased by the previous process of lemmatization and tagging of the text.[5] For this purpose, the text was

---

5 This text belongs to the lemmatized and tagged corpus of Middle English scientific prose compiled at the University of Málaga in collaboration with the Universities of Glasgow and Oviedo through the research projects mentioned above.

first transcribed directly from the digitized images of the manuscript. Then, a slightly modified transcription (in which word-division had been regularized, in the sense that hyphenation and truncation had been eliminated) was turned into an Excel file, where the lemmatization and tagging were carried out. On account of the presence of other Middle English scientific texts in the corpus and the great deal of orthographic variation in this period, the criterion of using the most frequent word was kept at bay so as to prevent a dialectally-biased database, among other shortcomings. Thus, the lemmas were taken from the entries in the *Middle English Dictionary*.[6]

Afterwards, all the items were tagged, including the word-class and accidence in each case (number, tense, degree, etc.). Though tiresome, this method has proved to be very useful in the analysis of dialectal provenance, inasmuch as all the spellings of a given item are retrieved promptly, for instance, all the verbs in the third person singular present indicative, or the various spellings of the item EYE in the plural. This, in turn, facilitates retrieving much more information. Accordingly, the spellings related to more than 70 different items of the questionnaire were initially retrieved.

4. Analysis of the dialectal provenance of the text

For a start, the language displayed in the text is, to a certain extent, quite standardized, which is typical of the Midlands at this time. However, the text exhibits a great number of variants for certain items, such as EYE (PL.), NOT, TOGETHER or IF.[7]

The questionnaire was filled in with items from the general section and from the sections where the northern and southern areas are surveyed. In both cases, functional words, inflexions and some content words have proved to be most useful. In the following table,

---

6 The online version can be found at <http://quod.lib.umich.edu/m/med/lookup.html>.

7 Following the conventions of *LALME*, items are systematically rendered in small caps.

the most salient items surveyed are presented (following the same order as shown in *LALME*), along with the variants collected and their corresponding frequencies (in parentheses). Only the most frequent variants of each item were taken into account for the dot maps:

Item	*Variants and frequencies*
THESE	*these* (11×), *this* (2×), *þese* (2×), *þes* (1×)
IT	*it* (407×), *hit* (18×), *itte* (1×)
SUCH	*suche* (30×), *soche* (13×), *such* (9×), *seche* (1×), *soch* (1×)
MANY	*many* (38×)
ARE	*ben* (80×), *be* (19×), *bene* (1×)
THOUGH	*though* (5×)
NOT	*nought* (86×), *not* (7×), *nouȝt* (5×), *naught* (3×), *nat* (2×), *noughte* (1×), *nowght* (1×)
THROUGH	
SUBSTANTIVE PL.	*-es* (480×), *-s* (84×), *-is* (77×), *-ys* (25×), *-us* (9×), *-ijs* (7×)
PRESENT PARTICIPLE	*-yng* (40×)
3RD SG. PRESENT INDICATIVE	*-ethe* (216×), *-ithe* (31×), *-eþ* (18×), *-ythe* (10×), *-eth* (4×), *-iþ* (3×), *-þe* (3×)
PRESENT INDICATIVE PL.	*-en* (120×), Ø (82×), *-yn* (26×), *-ethe* (5×), *-yng* (4×), *-un* (1×)
WEAK PAST PARTICIPLE	*-ed* (104×), *-id* (17×), *-yd* (10×), *-d* (4×), *-ied* (2×), *-it* (1×)
EYE (SG and PL.)	*eyȝe* (157×), *eyȝen* (136×), *eye* (6×), *eiȝe* (5×), *eiȝen* (3×), *ey* (3×), *yȝen* (2×), *enyn* (1×), *eyen* (1×), *eȝen* (1×), *yeȝen* (1×), *yȝe* (1×)
FLESH	*flesshe* (9×), *fflesshe* (2×), *flessh* (1×)
GIVE	*yeue* (9×), *yif* (3×), *yef* (1×)
SOME	*summe* (12×), *somme* (2×), *some* (1×), *sum* (1×)
SUN	*sonne* (7×), *sunne* (2×)
THREE (ORDINAL)	*third* (4×), *thirde* (4×), *thridde* (3×), *therd* (1×), *thirdde* (1×), *thride* (1×)
WITHOUT	*withouten* (4×), *withoute* (2×), *withouton* (1×), *withe oute* (1×), *with outen* (1×), *with outen* (1×), *with oute* (1×), *withowte* (1×)
YE	*ye* (91×)
YOU	*you* (24×)
-LY	*-liche* (42×), *-ly* (32×)

Table 1. Items searched for in the dot maps.

In view of the information obtained by means of the dot maps, the likely geographical location was restricted to the central and slightly northern part of the county of Warwickshire, probably more east than west. Other items which also point to this location are the *h*-forms in THEM and THEIR, the prefix <y–> (169×) for the past participle, *whan* (61×) (WHEN), etc. Hence, these results already confirm Eldredge's proposal as regards this vernacular prose text (1996: 44). However, not all of the variants of every item could be easily located in the dot maps, mainly as a result of the diverse stock of variants elicited for some items (chiefly IF, EYE, THROUGH, etc.). Moreover, some of the items checked against the dot maps were so general that they were of little use for determining a more specific geographical location, as in the cases of WHICH (all spellings ending in *-ch(e)*), or LITTLE (all spellings ending in *-ll(e)*). Therefore, the item maps were used in order to refine the results obtained. For this purpose, another, more restricted, questionnaire was devised, in which the following set of items was mapped:

Item	*Variants & frequencies*
WHICH	*the whiche* (37×), *þe whiche* (1×)
MUCH	*moche* (35×), *moch* (3×)
SHALL PL.	*shull* (45×), *shall* (7×), *schall* (1×), *shulle* (1×)
IF	*yif* (26×), *yef* (12×), *yf* (2×), *and* (1×)
AGAINST	*ayenst* (2×)
YET	*yit* (3×), *yet* (2×)
STRENGTH	*strength* (2×)
NOT	*nought* (86×), *not* (7×), *nouȝt* (5×), *naught* (3×), *nat* (2×)
THROUGH	*thorugh* (11×), *thorowe* (9×), *thorough* (2×), *thorow* (1×), *þorwe* (1×), *þourȝ* (1×)
EYE (PL.)	*eyȝen* (136×), *eiȝen* (3×), *yȝen* (2×)
FIRE	*fire* (5×), *fyre* (2×)
FIRST	*first* (38×), *ffirst* (13×), *firste* (5×), *ffirste* (2×)
LITTLE	*litill* (7×), *litell* (4×), *litil* (1×)
NEITHER	*nether* (1×), *noþer* (1×)
TOGETHER	*to giders* (5×), *to gidres* (4×), *to geder* (1×), *to gideres* (1×)
TWO	*two* (18×), *to* (3×), *twey* (1×)

Table 2. Items searched for in the item maps.

Despite the number of variants, the most frequent ones helped to further reduce the likely area of provenance to the south-western part of Leicestershire, which borders the north-eastern area of Warwickshire. In fact, most of the spellings can be located in the area surrounding LP 302 (a volume containing religious verse and prose) and 536 (a copy of Mirk's *Festial*) in Leicestershire, and 534 (a copy of the *Prick of Conscience*) and 699 (a copy of Malory's *Morte d'Arthur*) in Warwickshire. In this area, traces of both Northern and Southern LP conflate, although all of these LP belong to the Northern area. Nonetheless, some of these variants hint at the influence exerted by neighbouring areas, that is, central and southern Warwickshire (as shown in *þorwe* and *þourȝ, first(e)*; *eyȝen, eiȝen* and *yȝen*; *to gideres* and *twey*); west Northamptonshire (as in *thorough*); or northern Leicestershire (as in *nat*). This phenomenon is usually described as the normal dialectal variation which arises in border areas (McIntosh *et al*. 1986, 1: 13).

Finally, the less frequent variants (for instance, the remaining undetected variants of EYE [PL.], IF, or THROUGH, together with other items which are of dialectal significance and whose spellings were not recorded either in the dot maps or in the item maps) were searched for in the county dictionary, and grouped into roughly the same areas as mentioned above: a) south-west Leicestershire: *thoo* / *tho* / *þoo* (THOSE), *eny* (ANY), *yeff* (IF), *ayen* / *ayene* (AGAIN), *-eþ* / *-iþ* (3RD SG. PRESENT INDICATIVE), *eyen* (EYE) and *third* / *thridde* / *thirde* (THREE); b) Northamptonshire: *aȝeen* (AGAIN), *-ethe* (PRESENT INDICATIVE PL.), *eȝen* / *yeȝen* (EYE), *neither* (NEITHER), *thirdde* (THREE) and *to geders* (TOGETHER); c) north-eastern Leicestershire and south Lincolnshire *itte* (IT), *nowght* (NOT), *-eþ^e^* (3RD SG. PRESENT INDICATIVE), *therd* (THREE) and *to gidres* (TOGETHER); and d) south Warwickshire: *-ithe* (3RD SG. PRESENT INDICATIVE).

Yet, a few residual spellings are attested in areas completely alien to the proposed location, such as Sussex (SUBSTANTIVE PL. in -*ijs*), or either Shropshire or the Kent – Essex – Norfolk area (*thorwe*). Moreover, two of them (*seche* [SUCH] and *enyn* [EYE]) are located in Warwickshire, northern Leicestershire or Lancashire, and Shropshire or Lancashire, respectively. In this line, and although Lancashire is clearly far apart from the region proposed, there is an awkward term, *anȝoron* (2×), which could be considered to be characteristic of such area, if we assume that it is related to the modern plant name *andurion* (Grigson 1958: 375).[8]

5. Conclusions

On the grounds of the evidence available, we can argue that this is a vernacular text coming from the Midland area, and more specifically, from south Leicestershire, which confirms (and at the same time reduces) the dialectal area postulated by Eldredge. The mass of variants collected for some of the items suggests, though, that the language of the text is conflated with spellings from nearby areas, such as south Warwickshire, Northamptonshire, north-eastern Leicestershire and south Lincolnshire. A few spellings seem to come from other more remote counties, such as Shropshire or Lancashire.

Even though the number of variants may be extensive for some items, it is not possible to argue that this text is an instance of *Mischsprache*, as the influence exerted by other dialectal areas is limited, both in terms of the number of variants and of their frequencies. Minor variations are, however, detected from f.13 onwards, such as the frequent use of *-i-* in TOGETHER, or of *shall* as the plural form; the scarcity of *soch(e)* and of *-us* (SUBSTANTIVE PL.); the rise of *-lich(e)* as the adverb marker; the preference for *litill* instead of *litell*, or of *thri-* instead of *thir-*; as well as the use of *-yn* and *-ethe* (PRESENT INDICATIVE PL.). These differences, nonetheless, do not imply a major change in terms of the dialectal area where the text has been located.

Yet, the existence of forms completely alien to the county of Leicestershire, along with the various spellings for some items, have led us to consider the following hypothesis. It is not likely that these spellings can be entirely attributed to a different dialect in which the scribe might have been brought up, as they have been located in

8 The possible connection with the term *andurion* and Lancashire has been suggested by Andrew Breeze (personal communication).

several nearby counties. Instead, the fact that the scribe is acquainted with the language of such areas may suggest that he had become familiar with various spellings while travelling around the area. Another (and likelier) possibility is that he was a professional scribe who belonged to a scriptorium or 'house-style' and was, thus, used to handling documents of various provenances. In this line, Samuels first suggested the existence of a Central Midland Standard (1989), and Taavitsainen has recently studied the so-called 'Chauliac/*Rosarium* type', which appears in surgical and anatomical writings (2004: 214–215; 2005: 99–104).[9] The ophthalmological text under study shares some of the diagnostic features with the latter, as put forward by Taavitsainen (2005: 103), but the analysis carried out with the 'fit'-technique does not allow us to assign it with certainty to the 'Chauliac/*Rosarium* type'.

At any rate, dialectal studies such as this one show the increasing interest in Middle English scientific prose texts. Further conclusions about the process of transmission of this text, as well as the presence of certain spellings, can be eventually achieved by examining the relationship of Glasgow, University Library, MS Hunter 513 with the exemplar from which it is assumed to have been copied, Oxford, Bodleian Library, MS Ashmole 1468.

References

Benskin, Michael 1991. The 'fit'-technique explained. In Riddy, Felicity (ed.) *Regionalism in Late Medieval Manuscripts and Texts*. Cambridge: D. S. Brewer, 9–26.

Calle Martín, Javier 2000. Written Variation in *A Treatise on the Mensuration of Heights and Distances* (MS Sloane 213): An

9 Although its linguistic features can be located in the East Central Midlands, no medical school or scriptorium interested in the vernacularisation of Chauliac has been localised in this area, and this has led Taavitsainen to consider London as a possible location for the production of these texts (2005: 103–104).

Analysis of its Provenance. In Cortés Rodríguez, Francisco J. / González Orta, Marta / Hernández Pérez, Beatriz / Martín Díaz, M. Auxiliadora / Mele Marrero, Margarita / Pérez Quintero, M. Jesús (eds) *Variation and Variety in Middle English Language and Literature*. Barcelona: Kadle Books, 9–14.

Dossena, Marina / Lass, Roger 2004. Introduction. In Dossena, Marina / Lass, Roger (eds) *Methods and Data in English Historical Dialectology*. Bern: Peter Lang, 7–15.

Eldredge, Laurence M. (ed.) 1996. *Benvenutus Grassus. The Wonderful Art of the Eye. A Critical Edition of the Middle English Translation of his* De Probatissima Arte Oculorum. East Lansing, Michigan: Michigan State University Press.

Eldredge, Laurence M. 1998. A thirteenth-century ophthalmologist, Benvenutus Grassus: his treatise and its survival. *Journal of the Royal Society of Medicine*. 91 (January 1998), 47–52.

Eldredge, Laurence M. 2002. Did Benvenutus Grassus Lecture at Montpellier?. In Iyeri, Yoko / Connolly, Margaret (eds) *Essays on Medieval English presented to Professor Matsuji Tajima*. Tokyo: Kaibunsha, 155–167.

Esteban Segura, María Laura 2008. *G.U.L. MS Hunter 509 (ff.1r – 167v): An Edition and Philological Study*. Unpublished PhD Thesis. Málaga: University of Málaga.

Fernández, Francisco 1982. *Historia de la Lengua Inglesa*. Madrid: Gredos.

Grigson, Geoffrey 1958. *The Englishman's Flora*. London: Phoenix House.

Kurath, Hans / Kuhn, Sherman M. / Lewis, Robert E. / Reidy, John (eds) 1956-2001. *Middle English Dictionary*. Ann Arbour: University of Michigan Press. [Electronic dictionary: <http://quod.lib.umich.edu/ m/med/lookup.html>].

Laing, Margaret 1989. Introduction. In Laing, Margaret (ed.) *Middle English Dialectology: Essays on Some Principles and Problems*. Aberdeen: Aberdeen University Press, ix–xiv.

Marqués Aguado, Teresa 2008. *Edition and Philological Study of G.U.L. MS Hunter 513 (ff. 37v – 96v)*. Unpublished PhD Thesis. Málaga: University of Málaga.

Marqués Aguado, Teresa / Miranda García, Antonio / González y Fernández-Corugedo, Santiago 2008. *The Use of the Eyes: A*

Middle English Ophthalmic Treatise in G.U.L. MS Hunter 513 (ff. 1r – 37r). Málaga: Servicio de Publicaciones de la Universidad de Málaga.

McIntosh, Angus 1989a. A New Approach to Middle English Dialectology. In Laing, Margaret (ed.) *Middle English Dialectology: Essays on Some Principles and Problems*. Aberdeen: Aberdeen University Press, 22–31.

McIntosh, Angus 1989b. The Analysis of Written Middle English. In Laing, Margaret (ed.) *Middle English Dialectology: Essays on Some Principles and Problems*. Aberdeen: Aberdeen University Press, 1–21.

McIntosh, Angus / Samuels, Michael L. / Benskin, Michael 1986. *A Linguistic Atlas of Late Mediaeval English*. 4 vols. Aberdeen: Aberdeen University Press.

Milroy, James 1992. Middle English Dialectology. In Blake, Norman (ed.) *The Cambridge History of the English Language*, vol.2. Cambridge: Cambridge University Press, 156–206.

Moore, Samuel / Meech, Sanford B. / Whitehall, Harold 1935. Middle English Dialect Characteristics and Dialect Boundaries. *Essays and Studies in English and Comparative Literature. University of Michigan Publications in Language and Literature*. 13, 1–60.

Ogden, Margaret S. (ed.) 1971. *The Cyrurgie of Guy de Chauliac*, vol. 1. London, New York, Toronto: Oxford University Press.

Samuels, Michael L. 1983. Chaucer's spelling. In Grey, Douglas / Stanley E. G. (eds) *Middle English Studies: Presented to Norman Davis in Honour of his Seventieth Birthday*. Oxford: Clarendon Press, 17–37.

Samuels, Michael L. 1989. Some Applications of Middle English Dialectology. In Laing, Margaret (ed.) *Middle English Dialectology: Essays on Some Principles and Problems*. Aberdeen: Aberdeen University Press, 64–80.

Samuels, Michael L. / Smith, Jeremy J. 1981. The Language of Gower. *Neuphilologische Mitteilungen*. 82/3, 295–304.

Taavitsainen, Irma 2004. Scriptorial 'house-styles' and discourse communities. In Taavitsainen, Irma / Pahta, Päivi (eds) *Medical and Scientific Writing in Late Medieval English*. Cambridge: Cambridge University Press, 209–240.

Taavitsainen, Irma 2005. Standardisation, house styles, and the scope of variation in ME scientific writing. In Ritt, Nikolaus / Schendll, Herbert (eds) *Rethinking Middle English: Linguistic and Literary Approaches*. Frankfurt am Main, Berlin, Bern, Bruxelles, New York, Oxford, Wien: Peter Lang, 89–109.

Tabanelli, Mario (ed.) 1969. *Jehan Yperman, padre della chirurgia fiamminga*. Firenze: Leo S. Olschki.

Young, John / Aitken, P. Henderson 1908. *A Catalogue of the Manuscripts in the Library of the Hunterian Museum in the University of Glasgow. Planned and Begun by the Late John Young, Continued and Completed by P. Henderson Aitken.* Glasgow: Maclehose.

DAVID MORENO-OLALLA, ANTONIO MIRANDA-GARCÍA

An Annotated Corpus of Middle English Scientific Prose: Aims and Features[1]

1. Introduction

Tagged corpora can be said to fall roughly into two main classes: *labelled* and *annotated* ones. Some are labelled in the sense that the items of the corpus are added tags (normally as *headers* to particular sections of the text) to indicate just some, very general features (information like title, author, date, genre, and/or such macrotextual information as line and page). This permits easy identification and reference, which comes in particularly handy when you have to deal with mammoth corpora containing several million words. When employing this system, the customary procedure is that a string of items will share the same tag. For example, the words comprising the sentence To be, or not to be, that is the question could be tagged together under a single label, perhaps something like: <Shakespeare_Hamlet_III_i_56>, which indicates author, title of the work, act, scene and line, respectively. The following line of the prince's famous soliloquy, Whether 'tis nobler in the mind to suffer, would then offer something similar to <Shakespeare_Hamlet_III_i_57>. Further information can be added to this simple tag as needed regarding date of composition, *dramatis persona*, and the like. The actual syntax of the tag may vary exceedingly, depending on the mark-up language used; for example,

1 The present research has been funded by the Spanish Ministry of Science and Technology (grant number HUM2004–01075FILO) and by the Autonomous Government of Andalusia (grant number HUM–2609). These grants are hereby gratefully acknowledged.

using an XML syntax would entail some nesting (thus, line tags may be placed inside scene tags, which in turn are contained within act tags, etc.).

Annotated corpora, on the other hand, transcend the merely labelled ones in the sense that, besides the general and macrotextual information, they also provide data of linguistic nature (be it syntactical, morphological, dialectal etc.) tagged on the *individual* items of the corpus, rather than chunks of texts. Annotating the Shakespearean example just quoted would mean ten different tags, one for each word –although, of course, items *to* and *be* would probably be added the same tag on both occasions, at least if the corpus is morphologically or syntactically annotated.

It is plain to see why the latter kind of corpora is very much wished for in the academy. Results yielded from annotated corpora are richer, as they can provide us with vital information as to the usage of particular words, phrases and sentences (either isolated or in context), and allow us to retrieve extremely detailed information from any linguistic perspective: grammatical, semantic, pragmatic, sociolinguistic depending on the kind of tags added. Moreover, such detailed information can then be of further use if the output is inputted to computer-driven tools for, say, pattern-matching, automatic translation, authorship attribution, management of neural networks –not to mention simple search engines.

Unfortunately, annotated corpora also pose an important problem to the linguist, one that many find insurmountable. While indexing information like *author* or *title* into a raw text using labels is a straightforward and relatively quick task (even with enormous quantities of text), since it can be done semiautomatically using very simple macros in any word processor or XML editor, annotating all and every item of a text, even if only to record the simplest of information, is a toilsome, error-prone and time-consuming manual process that most people try to avoid. Hence it is hardly a cause of wonder that annotated corpora (be it morphologically, syntactically, discursively, etc.) are scarcer than one would hope although, fortunately, their number is increasing almost every year.

Scholars of Middle English are particularly fortunate in that there are both labelled and annotated corpora that cater for many of their needs. The most important labelled corpus is the *Corpus of Middle English Prose and Verse* (*Michigan Corpus* for short), which was compiled of the materials used for the *Middle English Dictionary*. Unfortunately the sheer bulk of works written in Middle English has meant that only a portion of the whole extant literature is available – unlike the *Electronic Corpus of the Dictionary of Old English*, or *Toronto Corpus* (di Paolo Healey 2004), which records virtually all extant Old English texts, tagged using HTML, SGML and XML formats. It consists of 146 complete texts, which were marked up using a SGML syntax and in conformance with TEI (*Text Encoding Initiative*) guidelines. The corpus can be freely accessed from the Internet.

The recent publication of the *Middle English Medical Texts* (Taavitsainen et al. 2005) has been an extremely welcome addition to the field of ME labelled corpus. This is a *second-generation corpus* in that it addresses a particular field of the ME language (Kytö 2006: 383), namely medical works during the period spanning 1375 to 1500. It includes 86 texts and almost half a million running words (exact figure 495,322) taken from both edited and unedited MSS. The corpus is best viewed using the *MEMT Presenter*, a programme written by Raymond Hickey that can also be used to make queries.

As for annotated corpora, by far the most substantial one is the second edition of the *Penn-Helsinki Parsed Corpus of Middle English* (acronym *PPCME2*; Kroch and Taylor 2005). This is a part of the *Penn Parsed Corpora of Historical English*, which also includes two corpora for Old English (prose and verse, respectively) and one for Early Modern English, all tagged using roughly the same codification. PPCME2 consists of either long excerpts or complete Middle English prose works, largely taken from the Middle English section of the *Helsinki Corpus of English Texts* (Rissanen and Ihalainen 1991). It contains almost 1.2 million running words (exact figure: 1,155,965). These have been morphologically tagged, syntactically parsed, and glossed. The CD-Rom is accompanied by an electronic tool (*CorpusSearch2*, written by Beth Randall), that can be used as a search program.

2. The *Annotated Corpus of Middle English Scientific Prose*

The *Annotated Corpus of Middle English Scientific Prose* (*ACMESP*) is a new annotated corpus that records several late Middle English *Fachprosa* treatises. The project is based at the University of Málaga, developed in cooperation with the Universities of Oviedo and Glasgow and financed by the Spanish Ministry of Education through a National project grant (HUM2004-01075/FILO). The aims of the National Project were not only to develop a corpus, but also to recreate the physical browsing experience of a MS in a virtual environment (hence, one akin to those contained in the British Library or the Wellcome Library websites, which were created using *Turning the Pages*, a commercial technology developed by Armadillo Systems), yet transcending the merely visual intentions of similar proposals by offering *a*) a physical description of the MSS, to make the visitors' experience all the more rewarding and informative through the coverage of some features that are not immediately evident in virtual environments (for example, measures, quiring or watermarks), and *b*) transcriptions of the texts, for the benefit of those untrained in mediaeval Palaeography. It was those transcriptions, added to the MSS images to ease reading, that were turned into a corpus that can be used to advantage, furthering therefore the purpose of the Project by allowing *c*) linguistic analysis. To this end, the corpus was annotated and then used as input to a computer application that makes searchable queries (including Boolean ones), and allows a statistical study of the inputted texts. The results of this process are readily available at the following web address: <http://hunter.filosofia.uma.es/manuscripts>.

In this corpus, the original ME texts have been approached from quite a different angle to that in Penn-Helsinki, even though we are confident that at least partial conversion to the tagging procedures used in PPCME2 could be achieved without experiencing many problems. The peculiar nature of its origins and its primarily visual interest explain that it has a much more limited scope: at present it contains about 300,000 running words, which have been morphologically tagged. On the other hand, this corpus has gained comparatively in internal coherence, since it addresses a very specific field (roughly, 15th century Medical texts in Middle English) and offers full texts, rather than excerpts. This is actually a valuable asset. Middle English *Fachliteratur* is in sore need of modern editions (excellent overviews of the current state of affairs can be found in Keiser 1998 and Voigts and Kurtz 2000), and hence is quite poorly catered by *PPCME2*, which relies on already published material only. In fact, only sections from Thornton's *Liber de Diversis Medicinis* and from a Berkshire *Treatise on Horses* can be found in the current edition of the corpus.

The *Malaga Corpus* (as we might call it for short) differs from PPCME2 in yet another respect: it is based on fresh examination of the actual Middle English MSS, all of them part of the Hunter Collection at Glasgow University Library, rather than on published material. This has allowed for the presentation of texts where editorial interference has been kept to a minimum: they offer, for example, the original line breaks and foliation/pagination, paragraph division and layout (presenting the scribal use of colours and *lemmata*, for instance) and, inasmuch as the current codification standards permit, also the original spelling and punctuation. Marginal annotations by people other than the main scribe, on the other hand, have been ignored in the understanding that these should be better regarded as different texts, although these, being an integral part of the artefact, have been duly transcribed for the benefit of the site visitors.

From a purely linguistic perspective, editorial invisibility can be of dramatic importance, as most paper editions are critical and therefore conflate readings from different MSS. This of course is fine if you are interested in knowing the contents of a text but is deplorable if your interest lies in dialectology as more often than not several layers of text will be silently blended together by the editor. Now, to which extent can we then positively state (as done in PPCME2) that this or that tradition is East Midlands, Southern, Northern? This can become a real problem if you are working with hundreds of different texts for, to which extent can you trust the dialect tags added to the items, if many of the texts are in fact modern, editorial composites?

This is avoided if every single text of a given tradition is tagged individually. Take for instance the case of MS Hunter 513, part of which houses a treatise on ophthalmology, *De probatissima arte oculorum*, which can also be found in MS Hunter 503. Its modern editor (Eldredge 1996), conflated both texts, using readings from MS 513 wherever MS 503 was defective. The problem is, MS Hunter 503 and MS Hunter 513 were demonstrably written in different parts of the country (the most salient phono-morphological features of Hunter 503 have been studied in Moreno Olalla 2006, while Teresa Marqués Aguado analyses the dialect of Hunter 513 elsewhere in this same volume).

3. Codification and Mark-up procedures

The several texts that form the Corpus have been codified using the Unicode standard (Allen and The Unicode Consortium 2007), in accordance with the current trend, so that the texts can be read using any Unicode-compliant typeface. We use *precomposed* forms (i.e. the character is formed by a single Unicode block: ā = U2101, ŏ = U014F), rather than *decomposed* ones (i.e. letterform + diacritic, not unlike the system used in a typewriter: ā = a, U0061 + ¯, U0304, ŏ = o, U006F + ˘, U0306), for the benefit of those using a dumb font for reading, since this type is much more extended among users, particularly those using an OS with an incomplete Unicode support. Unicode is very user-friendly (complex characters can be easily keyed in, with just simple adaptations of the keyboard most of the times), the value of the resulting graphs onscreen is clearly identifiable and it is in general less trouble-prone than using HTML syntax –not to mention the near-obsolete ASCII codification that the pioneering team at the University of Helsinki (Rissanen and Ihalainen 1991) had to turn to in the mid eighties, for lack of anything better at the time (exx: <+t> for thorn, <+d> for eth, etc.).

Mark-up has been done in two steps. The texts are first added headers, like in any labelled corpus, for identification purposes, and then the particular items are tagged. As for labelling, a header is included at the beginning of each new folio and then every five lines using the same syntax: `\HunterMSXXX (Section), folio/page, line\`. As for linguistic annotation, our model in the development of the tags was designed in such a way as to abide by the following requisites:

- It must be *universal*, meaning that the tags should be valid for any natural language.
- It must be *detailed*; the more information the tag gives, the more powerful the search engine will become as regards the retrieval of information, be it morphological, syntactic, lexical or dialectal.
- It must be *discreet*, allowing an easy counting of items or occurrences which fulfil the conditions set by the user through Boolean and non-Boolean filters.
- It must be *smart* in that it should allow the retrieval and analysis of information using both (running) words and lemmas.
- It must be *fast-readable*, both to machines and humans. Displaying the tags in tabular form is more convenient than as a running text, which usually becomes quite hard to the eye and demanding to the mind after a while, while arranging the information in a spreadsheet or as .php files allows for much more efficient processing than extracting information from a plain text file, since the search engine does not need to read the whole file sequentially over and over again, but just rearrange indexed columns and extract the information in a single stint.
- It must be *flexible*, in the sense that different morphological tags can be provided depending on the particular syntactic and semantic contexts. This also entails that the unit of analysis should transcend the simple word, using what Sinclair calls *lexical items* instead: 'one or more words which together realise a single unit of meaning' (Sinclair 2003: 5), thus including compound words and phrases.

Our interest lying primarily in morphology and dialectology, the tag has been developed accordingly, following usual POST (*Part-of-Speech Tagging*) procedures. The outcome therefore resembles very much the one used in other annotated corpora, even though we use a notation of our own. We yet diverge somewhat from the MED procedures in that a tag `Dete` is used within the field CLASS for cases like the article and the demonstrative, so as to be able to specify their syntactic function as constituent of a noun phrase. Note that MED tags most of these as pronouns, adjectives or adverbs. Similarly, the preposition *to* can be added a tag `RegAcc` within the field CASE/GOVERNMENT to indicate that this is the infinitive marker, rather than a 'proper' preposition. This (apparently minor) feature was added after we discovered that the presence or absence of such marker is to a given extent a stylistic choice which can be used in authorship attribution analyses. The variable has been applied, for instance, to decide which of the eighteenth-century *Disputed Federalist Papers* (Dawson 1863) were by James Madison, and which by Alexander Hamilton (Miranda García forthcoming).

We have occasionally split into two some tags that are understood as single entities in other corpora such as *PPCME2*. We have for instance segregated `Gradation` from `Adjective` and `Adverb`. The reason for this will become apparent if the policy employed by Kroch and Taylor for *PPCME2* is analysed. Penn-Helsinki covers the ground by resorting to six tags: `Adj` & `Adv` (positive), `Adjr` & `Advr` (comparative), and `Adjs` & `Advs` (superlative). The Málaga Corpus is able to display the same using just four: `Adje`, `Adve` (tags belonging to the field CLASS), `Comp` and `Sup` (belonging to the field SUBCLASS, or division to a class), which can be combined as needed. This is more time-efficient since the search engine is required to handle substantially less tags in the long run.

Tags are added to individual wordtypes using a spreadsheet. These were grouped into the following fields: LEMMA, CLASS, SUBCLASS, TYPE, SUBTYPE, GRADATION/TENSE, NUMBER, PERSON, CASE/GOVERNMENT, GENDER, PREFIX, LEXICAL and DIALECT/LANGUAGE. When using the spreadsheet, each field appears placed in different columns to the left of the wordtype, which are then filled with the appropriate tags. Note that most tags are hierarchically arranged. Thus, if a tag `Pron` is selected in the CLASS field, only those tags related to pronouns will be available at the SUBCLASS field, and so forth. To make searching as efficient as possible, accidents can coalesce when no clash seems possible, as with GRADATION/TENSE or CASE/GOVERNMENT.

Wordtype	Lemma	Class	SC	Type	Gr/Tn	Num	Per	Cs/Gv
These	this,r	Dete	Demo	*	*	Plur	*	*
bene	bēn,v	Verb	*	Anom	Infin	*	*	*
seuene	sēven(e,d	Dete	Nume	*	*	*	*	*
tonicles	tunicle,n	Noun	*	*	*	Plur	*	*
of	of,p	Prep	*	*	*	*	*	*
the	thē,d	Dete	*	*	*	*	*	*
eyʒe	eie,n	Noun	*	*	*	Sing	*	*
and	and,c	Conj	Copu	*	*	*	*	*
yif	if,c	Conj	Cond	*	*	*	*	*
the	thē,d	Dete	*	*	*	*	*	*
first	first,r	Pron	Nume	*	*	*	*	*
happethe	happen,v	Verb	*	*	PrsInd	Sing	3rd	*
to	tō,p	Prep	*	*	*	*	*	RegAcc
be	bēn,va	Verb	*	Anom	InflInfin	*	*	*
broke	brēken,v	Verb	*	*	PasPart	*	*	*
or	or,c	Conj	Disj	*	*	*	*	*
thorowe	thurgh,p	Prep	*	*	*	*	*	RegDat
any-thing	anīthing,r	Pron	Inde	*	*	*	*	*
ys	bēn,v	Verb	*	Anom	PrsInd	Sing	3rd	*
perforate	perfōrāte,a	Adje	*	*	*	*	*	*

Table 1. *ACMESP* in a spreadsheet environment (only a selection of fields is given).

Fields can be left blank during the annotation process if no tag applies (for example, PERSON if CLASS is `Noun`, or GENDER if CLASS bears the tag `Verb`). As a matter of fact, some fields will remain empty for the whole Middle English corpus, but they have been maintained in the grid because they are deemed relevant for other periods of the language. The clearest example of this is the field SUBTYPE, which serves mainly to distinguish classes within strong verbs and stems within strong nouns and consequently is of limited or no interest for the late Middle English period. An asterisk will fill the gap in these cases.

Although the meaning of most fields is self-explanatory, a brief description of some of them will be surely welcome. LEMMAS are normalised following the spelling system employed in the online version of the *Middle English Dictionary* (MED) whenever this is possible, or recreated following the same guidelines when the word is either not recorded in the dictionary (for instance, proper nouns like *Jerūsālem*, MS <ier*usal*em>; note the macrons following a normalized transcription of Hebrew יְרוּשָׁלַםִ)[2] or when there is some minor disagreement about the POS (for example, we tag *trŏubled* as an adjective, while MED chose to take it as past participle of *trŏublen* and has therefore no entry).[3] Diacritics indicating quantity (macron over long vowels, /u:/ frequently written <ŏu> –in opposition to the diphthongal <ou>) or quality (for example, palatal /g/ as <ğ>) have

2 No statement should be inferred from this concerning the actual phonetic quantity of the Hebrew vowels. Although most modern Hebraists are ready to agree in that the Tiberian system was one based on quality rather than quantity (see for instance Joüon 1996: § 6; Meyer 1966: § 11), the transcription of *šureq* (וּ) as <ū> and *qameṣ* (◌ָ) as <ā> in opposition to <u> for *qibbuṣ* (◌ֻ) and <a> for pataḥ (◌ַ) has stood as a convenient device from the very beginnings of Semitic Studies back in the Middle Ages (see W. Gesenius' preliminary remark on the matter, Kautzsch 1910: § 8), and still enjoys current use, particularly in introductory grammars (see for instance the table offered in Lambdin 1998: § 10). Other minor details have been disregarded as they had little value in a ME corpus: note for example the very broad transcription of *šewa* (◌ְ) as <e>, or the use of <s>, rather than <š>, for *šīn* (שׁ).

3 It is worth mentioning in this respect that MED not always followed its own policy: contrary to the case of *trŏubled*, a word such as *ashāmed* does have one entry as past participle of *ashāmen*.

been retained as they are given in MED. Occasionally though (for example, with words with an unclear etymology, as in the mysterious plant species called *annȝoroun*) the lemma reproduces the actual spelling in the MS.[4] Lemmas are followed by an abbreviation indicating part of speech, following the notation used in the MED. Even though this information will be repeated in the field CLASS, this was maintained as it is useful to separate homophones (for example, to distinguish the adjectival values of *much* from the corresponding adverbial ones), which is invaluable when making automatic queries involving lemmas.

The field DIALECT/LANGUAGE also deserves some comment. This is meant to contain tags related to dialectology. ME dialect studies have experienced a boost after the publication of the *Linguistic Atlas of Late Mediaeval English* (*LALME* for short; McIntosh, Samuels, and Benskin 1986). While reproducing the detailed analysis presented in the third volume of the Atlas (the 'County Dictionary') would have been the best policy or, at least, an extremely welcome addition to the tags, this was felt to make the tags belonging to this category extremely long and cumbersome at times –not to mention the probable infringement of current copyright laws.

We are therefore contented with providing just very broad locations, whenever there seems to be some *consensus doctorum* on the matter (for example, words with `a` representing the heir of OE *ā* will be marked *Northern*, or spellings such as `xal` as *East Anglia*), so it is fair to say that this field is actually left blank for many words and still awaits deeper work. But in any case this is a wholly different procedure from the one followed by the Penn-Helsinki team, since they use dialectal information as a header to a complete text rather than annotate individual lexical items as we do. As indicated above,

4 Professor Andrew Breeze from the University of Navarre has suggested (p.c.) that *annȝoroun* might be a misspelling for **andurion*. Would this prove true, it could then be argued that either the scribe or the lost exemplar of this MS might have originated in Lancashire, since this word is recorded as dialectal to this county to refer to Hemp Agrimony, *Eupatorium cannabinum* L. (Grigson 1987: 375, drawing from Wright 1896–1905: s.v.).

this can easily lead to erroneous assumptions when the text is a composite from different MSS or traditions.[5]

Problems in the annotation process were mainly posed by the LEMMA field. Reference has already been made to one of them: the lack of clear spelling standards during the period strongly suggested the convenience of adopting some spelling guidelines. Those used at the MED were used as they are the most authoritative lexical tool for ME and are available to anyone with an Internet connection free of charge. Some annotated corpora dealing with this period of the language (*PPCME2* for one) chose to dispense with them; but we think that the usage of univocal lemmas is a veritable help in this kind of corpora. This becomes tangible if, for instance, one were asked to separate which of the following forms belong to **sēn** 'to see' (< OE *sēon*, str. V) and which ones to **sein** 'to say' (< OE *seċġan*, wk. I): *saithe, say, say, saye, saye, sayen, sayen, sayne, sayne, saynige, se, see, seen, seethe, sein, seiþ, seithe, sey, seye, seyn, seyn, seythe, seythe, ysaide, ysee, ysey, yseyn*. *Seyn*, for instance, could easily be taken as the infinitive form 'to say', but it actually appears as the infinitive of the verb 'to see'. And these are forms taken from a single MS (Hunter 503)! One could argue that the context makes the meaning clear most of the times, but using an electronic corpus usually entails extracting single items out of the databases, detached most of the times from any clear context; there is where danger lies. Moreover and from a different perspective, lemmatisation is almost a prerequisite if you intend to tackle any study on vocabulary richness or authorship attribution studies.

The other issue posed by the lemmas is one which has to do with the actual definition of what the lemma might be, for example, in a phrasal verb. Should each component be taken in its own, or as a

5 An actual example of the dangers that scholars can be led to by taking such type of information for granted has been presented in Moreno Olalla 2007, particularly pp. 126–132. The manuscript analysed there (London, British Library, Sloane 5, ff. 13–57) is normally associated with Hereford, but a linguistic study proves that very little found there can be regarded as truly Herefordian (or, for all it matters, West Midlands): the text seems as a whole an East Midlands version, with a slight Northern layer at places that can be accounted for when the particulars of the MS are known.

whole? Each position will find defenders and corroborative arguments (and counterarguments). We have sometimes found it convenient to assign a single lemma to multi-word units, and sometimes we have decided to keep two or more lemmas where one could have been theoretically possible. The following list records the cases where we have used a single lemma for several items:

- in genitival compounds (*tatpuruṣa*), regardless of whether they are written as a single word, or graphically separated.[6]

```
a · corne milesede the whiche may \12r, 1\
were kyrnelle of myle sede wherfore my dere
\17v, 10\
```

- in scientific terminology, either in Latin, in English or in a mixture of both.

```
therwithe poudre of margerie stones and drye it
\35v, 15\
ffor that smytyng neruus opticus is stoppid in
\29r, 10\
that is cleped morsus galline · id est · \35v, 1\
```

- in combinations of function words when they evidently refer to a single function, like the following:

```
first yclepe Saluatricem for whi it sanethe all
\2v, 5\
```

- in combinations of function and content words which are on the brink of grammaticalization (Hopper and Traugott 2003).

```
the whiche ys be cause of smytyng and lyghtly
\6v, 15\
we sayne that thorugh accasioun of fleume \15r,
5\
```

- in relative/adverbial constructions with redundant that (including of course so that, which is still in use).

```
in that bacyn where that the fume of \13v, 1-4\
and whan that this is ydo \17r, 1-4\
```

6 For editorial convenience and since all examples quoted here belong to the same MS (Hunter 503), the tag has been reduced here to indicate just folio and line. The context is provided by the three items preceding and following.

consumed **first that** it seme that \11r, 10-14\
eyȝen bethe shitte **so that** he be naught \5v, 25\

On the other hand, phrasal and prepositional verbs have been split into their single components, since groups like cut her away \HunterMS513 (Eye), 22v, 20\ would pose problems to the lemmatisation syntax. Similar considerations have led us to tag comparative structures like *as... as* into two separate lemmas, each of them numbered for convenience. In instances like *aman*, *asmuch*, proclitics have been silently separated into a different lemma, even though the transcription has kept this type of scribal practices.

There have been issues of minor importance regarding other aspects of the process. They are briefly enumerated here, followed by their proposed solutions, if any.

- The first problem encountered was that of understanding the text correctly. It is not always the case that the scribe's meaning is made fully clear and syntactically awkward constructions, for instance, are very frequent in scientific texts. Punctuation is of little help, and can be particularly misguiding at times. Omissions and alteration of letters and words are half-expected, particularly when dealing with technical jargon, so that misspelt words are also very common. There is no easy remedy for all this, and mistakes on the taggers' comprehension of the ME text are almost doomed to happen, even more so since there is an ingredient of subjectivism involved in the process.
- When there is a clear mistake, the wordtype is kept as it stands in the MS, but it is given the correct lemma, as in <senwith> where the meaning intended was clearly **semith* 'it seems'. The lemma in this case is, therefore, *sēmen*.
- End of line division in the MS had to be attended to as well, in order to add the location tag. We decided that a word which runs through two lines would be understood as belonging to the first one. Thus, in

  and clepid it, Deus oculorum. Of the whi
  ch boke yn the fyrst Chapitre he declarith

  which is taken to belong to the same line as clepid or Deus.

- Another source of trouble was the tagging treatment of words from languages other than ME, like the *Deus oculorum* just mentioned. In the case of Latin wordtypes, lemmas have been taken from the standard *Oxford Latin Dictionary* (Glare 1997) whenever they are recorded in Classical Latin, and from the *Glossarium mediae et infirmae latinitatis* (du Fresne 1938) or the *Mediae latinitatis lexicon minus* (Niermeyer 1976) whenever they are Post-classical formations. Spelling for Greek words that were *not* first borrowed into Latin (and thus do not appear in the mentioned Latin dictionaries) but are found in Mediaeval treatises will be taken from the *Greek-English Lexicon* if from Classical Greek (Liddell and Scott 1996), or from the *Greek lexicon of the Roman and Byzantine periods* (Sophocles 2003) if not recorded there. French borrowings not recorded in MED will be Anglicized following the MED spelling guidelines or, if that proves impracticable, will appear as they stand in the *Dictionnaire de l'ancienne langue française* (Godefroy 1982). To avoid confusion, the field DIALECT/LANGUAGE will indicate that these are not 'proper' ME words, and note that no full morphological annotation is provided, as the value of the Latin/Greek morphemes can be misleading in these cases (see Moreno Olalla 2001: 157–159 on this point).

The final annotation looks like the following when invoked as a .txt file out of the spreadsheet:

```
These          [this, r:DeteDemoPlur<>{}]
bene           [bēn, v:VerbAnomPrsIndPlur3rd<>{}]
seuene         [sēven(e, d:DeteNume<>{}]
tonicles       [tunicle, n:NounPlur<>{}]
of             [of, p:PrepRegDat<>{}]
the            [thē, d:Dete<>{}]
eyȝe           [eie, n:NounSing<>{}]
and            [and, c:ConjCopu<>{}]
yif            [if, c:ConjCond<>{}]
the            [thē, d:Dete<>{}]
first          [first, r:PronNume<>{}]
happethe       [happen, v:VerbPrsIndSing3rd<>{}]
to             [tō, p:PrepRegAcc<>{}]
```

```
be                 [bēn, va:VerbAnomInflInfin<>{}]
broke              [brēken, v:VerbPasPart<>{}]
\MS513 2,R,15\
or                 [or, c:ConjDisj<>{}]
thorowe            [thurgh, p:PrepRegDat<>{}]
any thing          [anīthing, r:PronInde<>{}]
ys                 [bēn, v:VerbPrsIndSing3rd<>{}]
perforate          [perfōrāte, a:Adje<>{}]
and                [and, c:ConjCopu<>{}]
```

Annotating a corpus is no easy matter: editorial decisions are expected from the taggers, and it is only fair to admit that there is ample room for boredom and misery during the process. But in the long run annotation and lemmatisation have clear advantages, several of which have been mentioned in the previous pages, and that suggests the overall convenience of adding this information to the tags, even if that means devoting more time to this task. The profits are well worth the investment.

References

Allen, Julie D. / The Unicode Consortium 2007. *The Unicode standard 5.0*. Upper Saddle River, NJ: Addison-Wesley.

Armadillo Systems <www.armadillosystems.com/books.htm>.

British Library <www.bl.uk/onlinegallery/ttp/ttpbooks.html>.

Dawson, Henry B. (ed.) 1863. *The Federalist. A collection of essays, written in favor of the new Constitution, as agreed upon by the Federal Convention, September 17, 1787*. New York/London: C. Scribner - S. Low, Son & Co.

di Paolo Healey, Antonette 2004. *Dictionary of old English corpus*. (CDRom). Toronto: University of Toronto.

du Fresne, Charles 1938. *Glossarium mediae et infirmae latinitatis*. 10 vols. Paris: Librairie des Sciences et des Arts.

Eldredge, Lawrence M. (ed.) 1996. *Benvenutus Grassus. The wonderful art of the eye. A critical edition of the Middle English translation of his De probatissima arte oculorum*. East Lansing: Michigan State University Press.

Glare, Peter Geoffrey William (ed.) ^{2}1997. *Oxford Latin Dictionary*. Oxford: Clarendon Press.

Godefroy, Frédéric 1982. *Dictionnaire de l'ancienne langue française et de tout ses dialectes du IXe au XVe siècle: composé d'après le dépouillement de tous les plus importants documents manuscrits ou imprimés qui se trouvent dans les grandes bibliothèques de la France et de l'Europe et dans les principales archives départamentales municipales, hospitalières ou privées*. 10 vols. Genève/Paris: Slatkine.

Grigson, Geoffrey 1987. *The Englishman's flora: illustrated with woodcuts from sixteenth-century herbals*. London: Phoenix House.

Hopper, Paul John / Traugott, Elizabeth Closs ^{2}2003. *Grammaticalization*. Cambridge: Cambridge University Press.

Joüon, Paul ^{2}1996. *Grammaire de l'hébreu biblique*. Roma: Pontificio Instituto Biblico.

Kautzsch, Emil (ed.) ^{2}1910. *Gesenius' Hebrew grammar*. Oxford: Clarendon Press.

Keiser, George R. (ed.) 1998. *Works of science and information*. 10 vols. New Haven: The Connecticut Academy of Arts and Sciences.

Kroch, Anthony / Taylor, Ann 2005. *Penn-Helsinki Parsed Corpus of Middle English*. Pennsilvania: University of Pennsylvania.

Kytö, Merja 2006. Review of *New Frontiers of Corpus Research. Papers from the Twenty First International Conference on English Language Research on Computerized Corpora. Sydney 2000 (Language and Computers: Studies in Practical Linguistics 36). Literary and Linguistic Computing* 21/3, 382-384.

Lambdin, Thomas Oden 1998. *Introduction to Biblical Hebrew*. London: Darton, Longman and Todd.

Liddell, Henry George / Scott, Robert ^{9}1996. *A Greek English Lexicon. With a Revised Supplement*. Oxford: Clarendon Press.

McIntosh, Angus / Samuels, Michael L. / Benskin, Michael 1986. *A Linguistic Atlas of Late Mediaeval English*. 4 vols. Aberdeen: Aberdeen University Press.

Meyer, Rudolf [3]1966. *Hebräische Grammatik*. Berlin: Walter de Gruyter.

Middle English Dictionary <quod.lib.umich.edu/m/med>.

Miranda García, Antonio Forthcoming. The authorship of the 'Disputed Federalist Papers.' A lemma-based stylistic analysis.

Moreno Olalla, David 2001. Neologismos y errores léxicos en la traducción anglosajona del Evangelio según San Lucas. *Analecta Malacitana* 24/1, 153–173.

———*Yn dyuerse parties of the worlde*. Dialectal Blend in G.U.L. Hunter 503. Paper read at the the 18th International SELIM Conference, at Málaga, 2006.

——— 2007. *The fautys to amende*. On the interpretation of the *explicit* of Sloane 5, ff. 13-57, and related matters. *English Studies* 88/2, 119–142.

Niermeyer, Jan Frederik (ed.) 1976. *Mediae Latinitatis Lexicon Minus*. Leiden: Brill.

Rissanen, Matti Juhani / Ihalainen, Ossi Kalevi 1991. *The Helsinki Corpus of English Texts. Diachronic and Dialectal*. Helsinki: Helsingin yliopisto.

Sinclair, John McHardy 2004. Language and computing, past and present. Paper read at the *14th European Symposium on Language for Special Purposes*, (Guildford, 2003). Ed. K. Ahmad / M. Rogers / G. Anderman / D. Cheng / B. Antia (2004), 1–12; The paper is available online at: <www.computing.surrey.ac.uk/lsp2003/LSP%20PROCEEDINGS.pdf>.

Sophocles, Evangelinus Apostolides 2003. *Greek lexicon of the Roman and Byzantine periods (from B. C. 146 to A. D. 1100)*. Hildesheim - Zürich - New York: Georg Olms.

Taavitsainen, Irma / Pahta, Päivi / Mäkinen, Martti / Hickey, Raymond 2005. *Middle English medical texts*. Amsterdam: John Benjamins.

Text Encoding Initiative <www.tei-c.org/guidelines>.

Voigts, Linda Erhsam / Kurtz, Patricia Deery *Scientific and medical writings in Old and Middle English: An electronic reference* (ver. 1.0). University of Michigan Press, Ann Arbor, 2000.

Wellcome Library <library.wellcome.ac.uk/ttp.html>.

Wright, Joseph (ed.) 1896–1905. *The English dialect dictionary*. 6 vols. London: Henry Frowde.

NADIA OBEGI-GALLARDO

Lexical study of *The Book of Operation* (G.U.L. MS Hunter 95 (T.4.12) ff. 82^r-156^v): illnesses and treatment.[1]

1. Introduction

The treatise under scrutiny is a medieval scientific text entitled *Book of Operation* (hereafter referred as *BO*), housed in G.U.L. MS Hunter 95 (T. 4. 12). The first contact with this treatise took place in Glasgow as a result of a research period in that university where the edition of the *BO* began. After having worked with the manuscript itself at the Hunterian Library, a microfilmed copy was used and, in the last stage of the study, the digital edition allowed the analysis of the volume in more detail and in an easier way. In this case, a lexical study of the text is provided, which focuses on the field of illnesses and remedies, and the specific vocabulary employed during surgical operations.

Therefore, in light of these issues, the aims of this paper are the following: a) to classify the remedies mentioned in *BO*, mostly plant names, into compound or simple nouns and, more specifically, analyze the etimology and permanence of these words through time together with their collocations; b) to provide a classification of the illnesses mentioned in the treatise, attending to their survival until Present Day English (PDE); c) to study those verbs which help the scribe explain the surgical operations and other ways of healing; and d) to explain the main surgical instruments and treatments which are

---

1 The present study has been funded by the Spanish Ministry of Science and Technology (grant number HUM2004-01075FILO) and by the Autonomous Government of Andalusia (grant number HUM-2609). These grants are hereby gratefully acknowledged.

employed in *BO*. Thus, the contents are organized as follows: in Section 2 the methodology followed to carry out this study is explained; section 3 presents our analysis of the remedies; Section 4, in turn, will be devoted to the information with respect to the different illnesses found in the treatise, while Section 5 focuses on the selected verbs; section 6 deals with the surgical instruments and treatment and, finally, in section 7 the conclusions extracted from this study are drawn.

2. Methodology

Once the transcription of *BO* was finished, all the words of the text were listed and, subsequently, classified into their corresponding word classes, that is, nouns, verbs, adjectives, adverbs, determiners, and conjunctions by using an Excel spread-sheet.[2] Secondly, in reference to the nouns, a semantic classification allowed the extraction of those terms referring to remedies (320 terms), illnesses (90 terms), and intruments and treatment (40 terms). In addition, a further classification was carried out within remedies, by distinguishing between plants, on the one hand, and other products from nature or from animal origin, on the other.

3. Remedies

A total of 320 terms have been analysed, whether plants names or ingredients both from natural or animal origin. Before focusing on each of these groups, the method of classification in this particular case needs to be explained. Once all these words were included in an

2 All the headwords included in this analysis have been taken from the *Middle English Dictionary* in the website <http://ets.umdl.umich.edu/m/med/>.

Excel spread-sheet and arranged in alphabetical order, we looked for their cognates in PDE and Spanish, by consulting several PhD theses and research articles which deal with the same subject (Moreno Olalla, 2002; Norri, 1992; Peters, 2004, among others). At a first stage, we make a distinction concerning, on the one hand, those terms which have become obsolete (34.3%) and, on the other, those which have survived. Within the first group we find, for instance, *brassik* (cabbage), *yreos* (white flag), and *āche* (wild celery). Within the group of those which may still be found in PDE we find 65.7%: *acācia* (acacia), *affodil* (asphodel), *alōe* (aloe), *anēt* (dill), *antimōnie* (antimony), *anīs* (anise), *barlī* (barly), *betonike* (betony), *calaminte* (calamint), *camomille* (chamomile), *celidoine* (celandine), *cŏmīn* (cumin), *culrāğe* (culrage), *valeriān*(e) (valerian), *lupīne* (lupin), and *opopanak* (opoponax), etc.

Secondly, concerning word formation, we find simple or compound nouns. With respect to compound nouns, a total of 45 terms have been found in *BO*, but only some of them will be mentioned like *hōre-hŏune* (white horehound), *night-shāde* (black nightshade), *pōme garnate* (pomegranate), *san-dragŏun* (dragon's blood), *sŏur-dough* (sour dock), *tūt-saine* (tutsan), and *wal-wort* (danewort).

Thirdly, the etymology of each word was studied so as to verify the predominance of Germanic or Latin terms. On the one hand, the data extracted confirm that only 15.4% comes from Germanic, such as *asshe* (ash), *cōrn* (corn), *dile* (dill), *netle* (nettle), *nŏte* (nut), and *rīe* (rye). However, 84.6% of the words comes from Latin via Greek like *agarike* (purging agaric); from Old French (OF) and Latin such as *egremonie* (agrymony), *cardamome* (cardamos), and *ğenciāne* (official gentian); and from OF like *alma*(u)*nde* (almond), *baie* (bay), *baume* (balm of gilead), and *borāğe* (borage).

In addition, as mentioned above, we find not only plant names but also ingredients of natural origin like *butere* (butter), *clei* (clay), *vīne* (wine), *farīna* (flour), *sugre* (sugar), *sulphur* (sulphur), *vinegre* (vinegar), and *wax* (wax). Furthermore, ingredients of animal origin are widely used in most treatments like *fatnesse* (fatness), which may be of *cāpŏun* (castrated cock), of *gandre* (gander), of *lōmb* (lamb), of *porke* (pork), and *bācŏun* (bacon).

Another stage of this study deals with the analysis of the context where remedies are found, by taking into account both their function, which may be single or multiple, that is, they can be applied only to a particular remedy or, on the contrary, they are used in several remedies for different illnesses, and the part of the plant which is required for a particular treatment, whether the seed, the flower, the oil, etc. Within the group of those remedies with a single function, we have found the following, all of them to relieve pain: *antrax*, *apium*, *centorīe*, *epithīme*, *mirre*, and *smāl-āche*. However, there are other remedies which play a multiple function, as they can be used both to alleviate pain and to heal an infection or a specific illness, for example, *corāl* (for pain in the mouth and to restore the teeth which have been lost), *dile* (for an intestinal infection or stomach ache), *flammule* and *galbane* (both to alleviate pain and to heal the bite of an animal), *malwe* and *grāpe* (both to stop an haemorrhage and for internal pains), and *mul-berī*(e) (for infection in general and to soothe a sore throat). With reference to the particular parts of a plant which may be convenient for a specific treatment, we have found that one should take the seed of *mustard*, *kirnel*, and *chiche*; the flower of *fēnigrēk*, *lupīne*, *orobus*, *rīe*, and *whēte*; the oil of *alma*(u)*nde*, *anēt*, and *spīkenard*; the powder of *affodil* and *hermodactiles*; the water of *celidoine*, *hŏnī*, *tēsel*, and *verveine*; the root of *fenel*; the leaves of *minte* and *pimpernel*(e); the shell of *nŏte* and, surprisingly, we observe that *camomille* is the only plant from which the powder, the oil and the flower are used for different treatments.

Finally, it is also relevant to point out some collocations which specify a particular remedy; to be precise: adjective + noun collocations, for instance, *blak eleborin* vs. *white eleborin*; *white cerūse*; *drie figge*; *whyte lēd* vs. *rede lēd*; *coaguled milk*; *white popī*(e); *drie prūne* vs. *grene prūne*; *stronge vinegre* vs. *litel vinegre* vs. *hote vinegre*; *schewed whēte* vs. *good whēte*; *white wax* vs. *blake wax*; and *whik silvēr*.

4. Illnesses

Once the illnesses mentioned in *BO* were listed, they were classified depending on the part of the body which was affected, whether external or internal, and on their permanence through time, that is, if they have survived or not in PDE. Among the 90 terms studied in this semantic field, only 25 terms have become obsolete, whether the word itself or the specific meaning of *illness*. With respect to the second case, we find *brisūre* which, as a variant of *brusure*, disappears with the meaning of contusion or fracture; *calefacciŏun*, which only maintains in PDE the meaning of heated condition and not the medical or scientific sense; *pōre* does not survive or is rare as a porosity; *punctūre* does not maintain the meaning of pain that we find in *BO*; *resolūciŏun* as a dispersion of morbid matter in the body has disappeared; and, finally, *whītnesse* survives with the meaning of the quality of being white but not with the sense of lightness or fairness of complexion.

Furthermore, with reference to those names of illnesses which can still be found in PDE, we have found 65 terms such as *callōsitē* (callosity), *dropesī*(e) (dropsy), *erisipila* (erysipelas), *hernia* (hernia), *impetīgo* (impetigo), *sclīrōsis* (sclerosis), *tūberōsitē* (tuberosity), *ulcerāciŏun* (ulceration), *mātūrāciŏun* (maturation), *fēblenesse* (feebleness), *rānula* (ranula), and *squinacī*(e) (squinsy). To finish with the illnesses, two compound nouns which also appear in *BO* should be mentioned: the Latin term *malum mortuum* (malemorte), which has not survived, and *nōlī mē tangere*, which survives maintaining the same form, meaning any of various conditions causing ulceration of the skin.

5. Verbs

The verbs of the treatise under scrutiny are going to be analysed, particularly those related to surgical operations. Firstly, it should be mentioned that 68.3% of the verbs with a specific meaning which have been analysed comes from French or Latinate etymon, in contrast to 31.7% coming from Germanic, and they do not need any other word in order to specify a particular sense.

Within this group we find, among many others, *asswāğen* (alleviate, relieve), *cicatrīzen* (to heal by forming a cicatrix), *consouden* (to heal a wound or fracture by promoting cicatrization), *constreinen* (to thicken bodily fluids so as to diminish their flow or discharge), *enointen* (to apply a healing unguent or medicine to a wound), *garsen* (to make cuts or incisions in somebody so as to drain off venom or let blood), *mundifīen* (to cleanse a wound or infection), and *staunchen* (to stop blood from flowing). On the contrary, most of the verbs of general use come from Germanic and they are usually followed by a preposition or adverb to clarify the meaning. Within this second group we find *put upp*, *reysed upp*, *perteinen to*, *lyfte uppe*, *goeþ upp*, *bynde* (*it*) *uppe*, *leie on*, *put on*, *set* (*it*) *on*, *drope doune*, *passe doun*, *falleþ doune*, etc.

6. Instruments and treatment

After having talked about remedies, illnesses and verbs, we will focus next on those surgical instruments mentioned in *BO*; to be precise, only those which have proved to be more relevant in this research. For example, the *plūmāciol*, which is a soft pad of wool or cotton mostly used on wounds which presents different sizes, obviously depending on the size of the wound to be healed, as can be seen in the following figure taken from an 18th-century book of surgery:

36 DES OPERATIONS DE CHIRURGIE,

IV. FIG. DES PLUMACEAUX.

QUAND après une opération la plaie demande une tente ou une canule, on y en met une de celles que je viens de vous faire voir ; mais dans les plaies où il n'en faut point, on se sert alors de bourdonnets, qui sont des tampons de charpie, dont on remplit les cavités, & de plumaceaux dont on les couvre.

D'où vient le mot de plumaceau.

Le mot de plumaceau prend son origine de ce que les Anciens se servoient de plumes cousues en-

Figure 1. The *plūmāciol*.

Other instruments found in *BO* are the *carpīe* (a shredded linen, a surgical lint), the *clistēr* (a sort of syringe), the *girdel* (a belt or cincture used to cure hernia or hold medicines against the body, among other uses), the *lenticulār* (a surgical knife used in treating head wounds), the *levatōrī*(e) (an instrument, usually with an angled blade, used in trepanning and other operations to relieve pressure, free bone fragments, etc.), the *magdalēōn* (a small roll of plaster), the *rāsŏure* (to shave or cut hair), a *tente* (a roll of some soft material, sometimes medicated, which was placed in or near a wound to keep it open while it heals), the *ventōse* (used for drawing blood or other matter out of the body), the *mailet* (a hammer or mallet), the *pinsŏuns* (some sort of forceps used in surgical procedures, especially in tooth extraction), the *roinŏur*(e) (a cutting instrument, especially used in cutting holes in the cranium), the *roller*(e) (a roll of bandage), the *sclīce* (a flat cutting instrument or forceps); and finally, the *tāstŏur* (a surgical probe).

As a kind of summary, Figure 2 shows those medical instruments which were common in most of the surgical operations of the time. For instance, we observe several scissors, depending on the hardness of what was going to be cut, in the particular case of this treatise, hard or soft tissue of the skin; the same takes place with the different types of knives. In addition, some razor blades can be observed, that at the bottom left being the one which is used in the operations described in *BO* according to the brief description provided by the scribe.

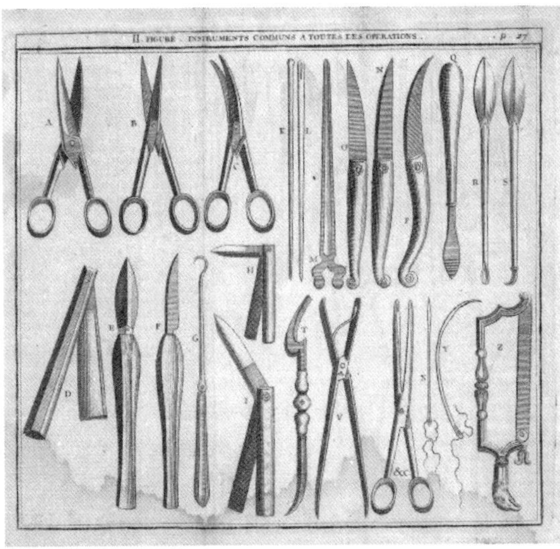

Figure 2. Surgical instruments.

It should be mentioned as well that in all the operations explained in *BO* there is a constant reference to compresses and emplasters (a salvelike preparation, usually of thick, viscous consistency, used in the form of a plaster). See, for example, the pictures of several emplasters and compresses of different sizes, once again depending on the area which is going to be treated:

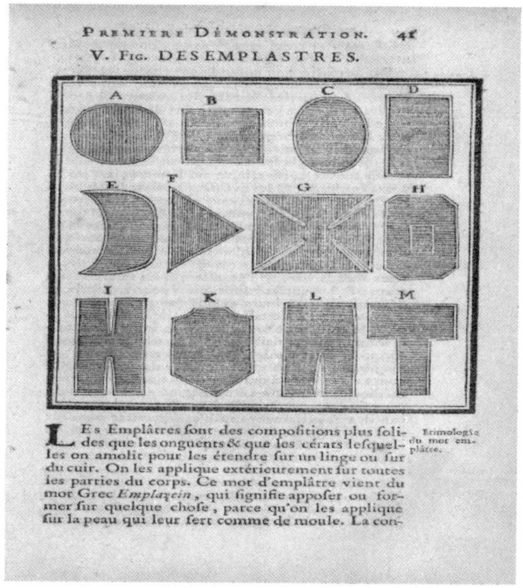
PREMIERE DÉMONSTRATION. 41

V. FIG. DES EMPLASTRES.

LEs Emplâtres font des compofitions plus folides que les onguents & que les cérats lefquelles on amolit pour les étendre fur un linge ou fur du cuir. On les applique extérieurement fur toutes les parties du corps. Ce mot d'emplâtre vient du mot Grec *Emplazein*, qui fignifie appofer ou former fur quelque chofe, parce qu'on les applique fur la peau qui leur fert comme de moule. La con- Etimologie du mot emplâtre.

Figure 3. Emplasters.

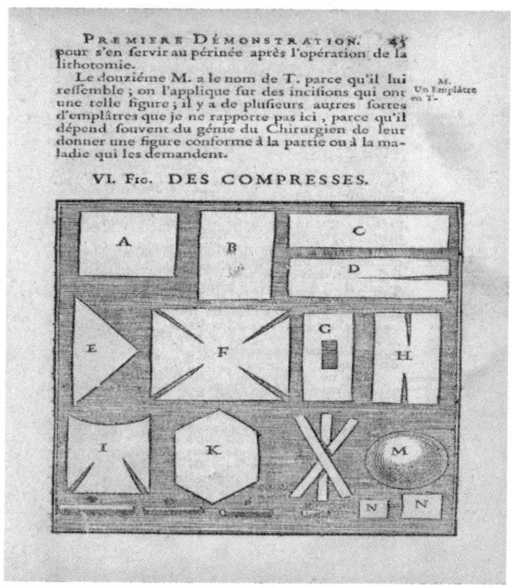
PREMIERE DÉMONSTRATION. 45

pour s'en fervir au périnée après l'opération de la lithotomie.

Le douziéme M. a le nom de T. parce qu'il lui reffemble ; on l'applique fur des incifions qui ont une telle figure ; il y a de plufieurs autres fortes d'emplâtres que je ne rapporte pas ici, parce qu'il dépend fouvent du génie du Chirurgien de leur donner une figure conforme à la partie ou à la maladie qui les demandent. M. Un Emplâtre en T.

VI. FIG. DES COMPRESSES.

Figure 4. Compresses.

Finally, we refer to the *terciān* which is also mentioned in the treatise, a container for liquids holding aproximatelly 84 gallons, used when mixing different ingredients for a specific cure. This leads us to talk about treatment, that is, the ordinary methods described in *BO* and which are present in most of the remedies, to be precise, 60 terms have been studied but we will deal only with the most significant ones. On the one hand, we group those which need heat to be effective such as the *triafarmacum* (a hot plaster made of wheat flour, water, and oil), the *dēcocciŏun* (a medicinal preparation made by boiling), and *fōmentāciŏun* (the process of applying hot medicinal substances to the outside of the body). On the other, there are some treatments which have to be composed of plants or other natural ingredients, such as the *cathaplasma* (a *poultice* containing herbs, in the case of *BO* against any illness in bones), the *mūscilāğe* (a thick, gelatinous substance obtained from the seeds or roots of certain plants); the *grūel* (a poultice, especially one containing meal), and the *pultes* (in this case, the so-called maturatives which induce suppuration).

In addition, many other treatments are explained by the scribe which involve the cleansing or purification of the part of the body to be treated as in the processes of *lavāciŏun* and *mundificāciŏun*. Moreover, there are several types of ointments such as the *dyachilon* or *nūtrītum*, in the case of *BO* against ulcers and wounds in general. Other treatments involve the drying of a specific area to prevent an infection as *dēsiccāciŏun* and *exsiccāciŏun*, or the processes of *epithima* and *strictory* which need a compress or cloth to heal an external wound. Finally, we mention more specialized tretaments, namely, *cochie* (a remedy for relieving excess humors), *excōriāciŏun* (the removal or cutting away of skin), *repercussiŏun* (the action of driving back humors from a swelling), *resolūciŏun* (the dispersal of an infection), *restōrāciŏun* (the mending of a broken bone, fracture, or dislocation), *scarīficāciŏun* (the making of shallow incisions in a part of the body as a means of letting blood), *sigillāciŏun* (sealing a wound), and *trōciske* (a medicated tablet or pellet).

7. Conclusions

Once all the data considered for this study have been analysed, we can draw the following conclusions. Concerning remedies, we have proved that most of them (65.7%) are still being used as natural treatments, thus making easier the task of understanding the methods of surgery in the Middle Ages. Moreover, it has also been relevant to prove that the great part of the remedies fulfil a multiple function as they are involved in most operations in *BO*. As far as illnesses are concerned, only 25 terms have become obsolete, which reveals that most of the scientific vocabulary comes from Latin. Most verbs, on their part, and just like remedies, come from OF and Latin and, due to this fact, they do not need any preposition or adverb to specify a particular sense. Finally, we also highlight the supremacy of Latin etymology in the case of the names designating instruments and treatment, as only 9.6% of these words comes from Germanic.

References

Bosworth, Joseph / Toller Thomas Northcote. 1991. *An Anglo-Saxon Dictionary*. New York: Oxford University Press.

Clark-Hall, John Richard. 1996. *A Concise Anglo-Saxon Dictionary*. Toronto-London: University of Toronto.

Esteban Segura, Laura. 2006. *Léxico y terminología en G.U.L. MS Hunter 509*. Masters Diploma of Advanced Studies. University of Málaga.

Getz, Faye Marie. 1998. *Medicine in the English Middle Ages*. Princeton, N.J.: Princeton University Press.

Kurath, Hans *et al.* 1959-. *Middle English Dictionary*. <http://ets.umdl.umich.edu/m/med/>.

Moreno Olalla, David. 2002. *Lelamour's Herbal (MS Sloane 5, ff. 13-57): Critical Edition and Philological Study*. Unpublished PhD Thesis, Universidad de Málaga.

Norri, Juhani. 2004. Entrances and Exits in England Medical Vocabulary, 1400-1550. In Taatvisainen, Irma / Pahta, Päivi (eds) *Medical and Scientific Writing in Late Medieval English*. Cambridge: Cambridge University Press, 100-143.

—— 1992. *Names of Sicknesses in English, 1400-1550: an Exploration of the Lexical Field.* PhD Thesis, Helsinki, Suomalainen Tiedeakatemia.

Obegi Gallardo, Nadia. 2006. *Estudio léxico-semántico, diacrónico y traductológico de términos de enfermedades y dolores en inglés antiguo.* Masters Diploma of Advanced Studies. University of Málaga.

Oxford English Dictionary. 1999. CD-Rom version 2.00. Copyright © Oxford University Press. <www.oed.com>.

Peters, Hans. 2004. The Vocabulary of Pain. In Kay, Christian, J. / Smith, Jeremy J. (eds) *Categorization in the History of English*. Amsterdam: John Benjamins, 193-220.

Segura Munguía, Santiago. 2004. *Diccionario etimológico de medicina.* Bilbao: University of Deusto.

Stratmann, Francis Henry. 1994. *A Middle-English Dictionary*. New York: Oxford University Press.

The International Plant Names Index. 2004. Royal Botanic Gardens, Kew, Harvard University Herbaria and Australian National Herbarium. <http://www.ipni.org>.

Vriend De, Hubert Jan. 1984. *The Old English Herbarium and Medicina de Quadrupedibus*. London-New York-Toronto: Oxford University Press.

ELENA QUINTANA-TOLEDO

Orality in the Middle English Medical Recipes of G.U.L. Hunter 185

1. Introduction

The study of Middle English medical recipes has increased considerably in the last two decades, not only from a purely philological standpoint but also from a wider linguistic perspective. The former includes editorial, palaeographical and codicological works, whereas the latter has focused on analyses in terms of text-type features and genre, as well as on pragmatic accounts, among others. Nevertheless, little attention has been paid to the extent to which Middle English medical recipes have preserved traces of orality, if any at all, and even less research has been carried out following a well-defined theoretical framework.

Remedy books have been frequently considered to be the result of a popular tradition in contrast with academic books whose origins tend to be traced back to a more learned one. Moreover, their intended audience consisted primarily of lay people (Voigts 1982), even though there seemed to be some constraints such as the ones pointed out by Alonso-Almeida and Carroll (2004) regarding literacy and economic factors. Their often-alleged popular origin and audience could have rendered some traits of orality to them, which would make them a rather interesting corpus of study in this sense. Obviously, I do not mean to equate the terms *oral* and *popular* since there is no conclusive evidence that learned cultural manifestations are void of oral features. However, supposed folk medical texts can be an appealing starting point in order to ascertain whether or not there is some sort of connection between orality and popular tradition/audience.

This paper seeks to explore the degree of orality which may have been transferred from spoken language to a corpus of medical recipes excerpted from the fifteenth-century manuscript G.U.L. Hunter 185, an example of a medieval remedy book. The search for data will rely on Ong's characterisation of orally based expression as specified in his seminal work *Orality and Literacy. The Technologizing of the Word* (1982) as it stands as one of the few texts that offers a fairly detailed description of oral style features, though from a macro level viewpoint. I shall try, then, to provide a micro level analysis by concentrating on their possible pragmatic interpretation.

This said, the paper is structured as follows: the next section focuses on the description of the texts that will serve as the corpus of study. They are medieval English medical recipes which have been excerpted from a remedy book. The third section holds some ideas related to the concept of orality and the theoretical framework, that is, Ong's proposal of features of orality. Special attention will be paid to the mingling of orality and literacy as it results in the occurrence of oral traits in the written mode. The fourth section is devoted to the analysis of the texts by searching for the data in relation to the characterisation proposed by Ong. Finally, the last section contains the conclusions that can be drawn from the present study.

2. Description of the texts

The texts analysed here consist of a collection of medical recipes belonging to the fifteenth-century remedy book contained in G.U.L Hunter 185 (henceforth H). They form a corpus of twenty thousand words approximately. It was edited by Alonso-Almeida (2000) and has been mentioned in some works such as Voigts and Kurtz (2000), Carroll (2003) or Taavitsainen, Pahta and Mäkinen (2005).

This manuscript is part of the *Corpus of Early English Recipes* (*CoER*), a larger research project which is currently underway in the

University of Las Palmas de Gran Canaria and is being carried out by the *TeLL* (*Tecnologías Emergentes Aplicadas a la Lengua y la Literatura*) research group in collaboration with members of *CeTIC* (*Centro Tecnológico para la Innovación en Comunicaciones*). *CoER* pursues a twofold aim: firstly, the compilation of Middle English and Renaissance recipes, which are typed directly from the original sources, and secondly, the design and implementation of an information retrieval tool which allows researchers to work with the texts. It is worth-mentioning that H also appears in the corpus of *Middle English Medical Texts* compiled by Taavitsainen, Pahta and Mäkinen.

H is made up of four items: the first one is a list of plants alphabetically arranged under their Latin names. The second item is a list of ingredients as used in medical preparations. The third one includes some notes regarding the medical properties of some natural products. And, finally, the last item is a collection of recipes together with some charms and prognostic texts. It is precisely this fourth item that will be scrutinised in the present study.

H appears to be an interesting corpus for a number of reasons: being a remedy book, it contains fairly straightforward directions to cure diseases whose structural organisation is one of *de capite ad pedem* order. Moreover, several types of reader-friendly devices are employed such as green or red colours to indicate the beginning of the remedies assisting the user in their reading. In addition to this, its small dimensions (183 x 89 mm) may suggest that it was to be easily taken from one place to another and frequently used. Lastly, the major parts are written in English, but French lexical items and Latin passages can be found, too, and so, it might be analysed from a code-switching perspective.

As regards the recipes proper, their internal structure tends to be divided into several sections (following the description offered in Alonso-Almeida (2000)): (i) title, which indicates what the remedy is for, (ii) ingredients, which names the substances that are needed in order to make it, (iii) preparation, which provides the instructions in relation to the combination of ingredients, (iv) application, which describes how the remedy is to be used, and (v) efficacy phrase, which offers a validation for the remedy in question.

3. Orality

Characterisations of orality range from those where categorical distinctions are made between orality and literacy to those where they cannot be easily distinguished from each other but rather the contrary. On the one hand, in the case of categorical distinctions the emphasis falls on the shift from a completely oral culture, where there is not any written manifestation whatsoever, to a literate culture, where writing is said to nearly supersede oral communication. On the other hand, views based on the notion of a mingling of orality and literacy do not place these two concepts at the two extremes of a pole: there is a continuum of cases and it may well be that written communication shows features of oralism, and that oral performance contains, to a certain extent, discourse patterns typically associated with writing.

Interest in oral traditions started in the first half of the twentieth century thanks to the research of scholars like Milman Parry (1971) and Albert Lord (1960) in the fields of Greek and Balkan literatures. However, it was not until the sixties when such studies experienced an increase in importance and a theory of orality began to be substantiated. Havelock (1991: 11), one of the founding figures, defines orality and oralism as:

> characteriz[ing] whole societies that have relied on oral communication without the use of writing. They also seek to identify a given type of language used in oral communication. And finally, they are used to identify a given type of consciousness, which is supposed to be created by orality or expressible in orality.

In his writings, he points out that in a preliterate Greek society poetry was the only way cultural knowledge could be kept and passed on from generation to generation (1977: 3), and that this could be achieved by means of (i) rhythmic patterns, which facilitated recalling (1963: 100), (ii) verbal formulas, whose repetition memory benefited from (1963: 82-84) and (iii) storytelling derived from the gathering of smaller stories by following an episodic order using specific agents as links (1963: 175-176). The status of poetry as a culture-preserving means changed with the advent of writing.

The breakdown occurred in ancient Greece with the invention of the alphabet. This was designed to represent the sounds of speech as closely as possible and, as Greek culture became alphabetised, the existing method for preserving knowledge, namely, the spoken word, started to be slowly superseded by a new way of storing information, that is, the written word. After a process of stabilisation, it was acknowledged that writing did not have some of the limitations associated with oralism such as the need for repetition to memorise information and, in turn, the need for memory in order to recall. Speech sounds, then, "could become a document, a permanent set of visible shapes, no longer a fleeting vibration in the air but shapes that could be laid aside until rescanned for some purpose and indeed forgotten" (1991: 25). In practice, the introduction of writing allowed writers to develop speculative thinking and particularly philosophical, prose discourse which Western civilisation has rested upon.

Plato unconsciously illustrates the shift in his Republic when he rejects epic poetry as it results in "a version of experience which is twice removed from reality" (1963: 3). Havelock notes that his harsh criticism does not simply imply a rejection of traditional poetry *per se*. Most importantly, it symbolises the rejection of the oral tradition from which poetry emerged as a vehicle for knowledge transmission. This shift is the main issue around which Havelock articulates his theory on Greek orality and literacy. Still, he does not overlook the possibility of both orality and literacy influencing each other. As a matter of fact, he (1991: 11) posits the following:

> The two, orality and literacy, are sharpened and focused against each other, yet can be seen as still interwoven in our own society. It is of course, a mistake to polarize these as mutually exclusive. Their relationship is one of mutual, creative tension, one that has both a historical dimension – as literate societies have emerged out of oralist ones – and a contemporary one – as we seek a deeper understanding of what literacy may mean to us as it is superimposed on an orality into which we were born and which governs so much of the normal give and take of daily life.

This view of orality and literacy mingling with each other is shared by other scholars. Tannen (1982) theorises about orality and literacy in terms of an oral/literate continuum instead of a clear differentiation between the two. She concludes that both spoken and written discourse can display clear features of oral and written style at the same time by virtue of the addresser's relative focus on his/her personal involvement with his/her addressee together with the specific context for each communicative situation.

Furthermore, she deepens into the study of oral and written discourse by the application of the theory of frames to a corpus of modern narrative texts in English and Greek (Tannen 1986). She identifies several frames on the basis of the contextual parameters which govern the communicative exchange, the content of the message conveyed and the linguistic evidence available. Some of them are the-subject-of-experiment frame, the oral storytelling frame and the film frame. In particular, I would like to highlight the oral storytelling frame since storytelling, as Havelock noted, is typical of preliterate, oral societies. This frame appears to model many written narratives and its presence is attested by some conventions in Tannen's data: (i) crossed-out lines, (ii) sense of immediacy, (iii) side sequences, (iv) formulaic expression and conventionalised sentence structure, (v) lexical choice associated with oral storytelling and, (vi) the use of deixis.

Apart from Havelock and Tannen, there is still another scholar whose contribution to studies into oral and literate traditions is relevant to the present paper, that is, Walter Ong. His relevance is determined by the fact that he offers a thorough description of orality features. This will serve as the theoretical framework for the search for data in the Middle English recipes of H. According to him (1982: 6), orality can be either primary or secondary. The former refers to the type of orality existing in cultures in preliterate times when knowledge survived and was passed on from generation to generation by word of mouth. In contrast, secondary orality belongs to literate cultures, all of which are based on the introduction of writing and print. Although primary and secondary orality seem to stand at opposite ends in the orality/literacy continuum, secondary orality makes it possible for some characteristics of primary orality to be embedded in writing.

Ong (1982: 36-57) points out that orally based expression is characterised by being: (i) additive rather than subordinative, (ii) aggregative rather than analytic, (iii) redundant or copious, (iv) conservative or traditionalist, (v) close to the human lifeworld, (vi) agonistically toned, (vii) empathetic and participatory rather than objectively distanced, (viii) homeostatic, and (ix) situational rather than abstract. I shall now comment on each of them briefly.

- Additive rather than subordinative: Expression is based on the addition of one chunk of information after another by means of avoiding any kind of subordinative relations between them.
- Aggregative rather than analytic: The constituents of oral expression "tend to be not so much simple integers as clusters of integers" (Ong 1982: 38). In practice, this means that orality makes frequent use of similar/dissimilar terms or phrases, epithets, formulas and clichés.
- Redundant or copious: Repetition is favoured in oral expression because it allows the audience to follow what has been uttered. It is a way of dealing with the fleeting quality of the utterances.
- Conservative or traditionalist: Orally transmitted knowledge must be repeated over and over again in order not to disappear. Repetition and memorisation tend to prevent experimentation from taking place.
- Close to the human lifeworld: Knowledge is conceptualized and verbalized with closer reference to the human lifeworld. There is a tendency for anything that is told to be passed through the sieve of familiar interaction of human beings.
- Agonistically toned: Oral expression seems to be pervaded by the idea of confrontation. Knowledge is often achieved after participants have become contenders in verbal or intellectual struggles.
- Empathetic and participatory rather than objectively distanced: Narrators, characters and audience are closely related to each other as key participants since at any time they might get involved in oral discourse.
- Homeostatic: Orality is anchored in the present. Word-meaning is determined by the contextual parameters which govern the

setting in which an oral discourse is being uttered. Real-life situations shape the meaning of the words here and now.
- Situational rather than abstract: Abstraction is kept to a minimum and situational frames of reference are dominant due to the aforementioned closeness to the human lifeworld and the homeostatic quality of orality.

For the purpose of the present paper, I have selected only some of these, namely, (i) additive rather than subordinative, (ii) aggregative rather than analytic, (iii) redundant or copious and (iv) close to the human lifeworld. The motivation behind this particular selection is that the characteristics of my corpus would yield imprecise data for features like conservative or traditionalist, which would be better looked at in a larger corpus covering several periods of the history of English. Furthermore, the rest of them appear to overlap with the ones I have chosen and so I have decided to somewhat simplify them.

4. Analysis

This section is devoted to the analysis of the data taken from H. In order to do so, I have made a division into subsections which correspond with some of the features of orality according to Ong's framework, that is, (i) additive rather than subordinative, (ii) aggregative rather than analytic, (iii) redundant or copious and (iv) close to the human lifeworld. I shall try to account for the data obtained by focusing on their pragmatic explanation. The texts are presented as they appear in the original, avoiding editorial intervention as much as possible. It should be noted, however, that the letter yogh has been substituted for 3.

4.1 Additive rather than subordinative

The written as opposed to the oral mode is traditionally characterised by its heavy reliance on subordination (Kroll 1977, Chafe 1982). According to Ong (1982: 38), the occurrence of subordinate, analytical discursive patterns is a way of providing meaning to the information conveyed through writing because it lacks the immediate context which shapes meaning in oral settings.

The texts analysed here do indeed contain a massive use of addition instead of subordination. Paratactic arrangements are dominant, being primarily realised by an overwhelming employment of the ampersand (&) at the expense of few instances of *and* to overtly mark clausal coordination. The following examples illustrate this point:

> (1) Also for
> scabbes in a mon or in a womman
> Tak þe rote of horshelne & seeþ it in
> water til hit be nessche & tak olde sche
> pes talwe & medle hem wel to gedre
> in a mortere & do it in a lynnen cloþ
> & anoynte þe sore agayn þe fuyre &
> he schal be hol

> (2) For the canker in mannes body • Tak
> þe rote of dragance • & schere it in small
> peces & let it dry3e • mak powdere þerof
> & ix peny wyght of þat powdere • & do it
> in hoot water al nyght & a morewe pou
> re owte þe water & do þerto whyt wyn &
> seth it wel & let þe seke drynke þerof warme
> & with jnne þre drykynges he schal be hool

Almost all of these two recipes are made up by a succession of clauses coordinated by the ampersand, which is employed to introduce a new informative chunk every time. *Et* fulfils the same function as the ampersand in those passages written in Latin, but together with other linkers such as *but* and *or* is kept to a minimum.

As has been shown in Alonso-Almeida (2008), *and*-conjunctives can convey meanings which depart from the one conveyed by the logical operator &. These meanings depend on the context in which *and* appears and can express a wide array of relations between the conjuncts involved like time, purpose or result, among others. The ampersand is mostly used in these examples with a temporal value. Consequently, the state of affairs described in the conjuncts must be understood as occurring in a temporal sequence. It should be noted that if the chronological relation is not followed and the sequencing of events is reversed, the remedy will not have the expected results.

As for the hypotactic arrangements, they are scanty in these recipes. Nonetheless, some of them can be found. That is the case of example (1), where the conjunction *til* is used to introduce a subordinate clause of time. Actually, the vast majority of the hypotactic constructions that appear in these recipes take on the grammatical function of adverbials of time in the main clause. *When* or *whanne* together with *til* are the subordinators used in these instances. Other less frequent subordinate clauses are conditionals, introduced by *ȝef* and *ȝif*, and relatives, introduced by *that*, *þat* and *whiche*. As for their position, while adverbial clauses of time and condition are mainly used in the preparation stage where the scribe indicates how the ingredients must be treated in order to achieve an efficient therapeutic remedy, relative clauses might be found at any stage.

On the whole, the elaboration of the grammatical patterns employed in these examples cannot be said to be complex, except for the presence of some adverbial (time and condition) and relative clauses as already pointed out. This suggests an oral residue, which becomes particularly noticeable when these recipes are read aloud.

4.2 Aggregative rather than analytic

Ong (1982: 38) posits that this feature of orally based expression is closely related to memory which, in turn, plays a vital role in performance. Human working memory at a given point during a speech is certainly constrained, so that not all chunks of information are equally accessible or even accessible at all. Formulaic language provides speakers with a mnemonic device which renders the recalling process easier since formulas tend to be formed by means of fixed and, sometimes, rhythmic patterns. The same would apply to epithets and clichés, though they are not present in the medical recipes analysed in this paper. As for the formulaic items, they display a high frequency of use.

Formulas are not necessarily individual lexical items. They may also include fixed expressions, idioms and clichés (Overstreet and Yule 2001: 3). Wray (2002: 9), for instance, defines a formulaic expression in an inclusive sense as

> a sequence, continuous or discontinuous, of words or other elements, which is, or appears to be, prefabricated; that is, stored and retrieved whole from memory at the time of use, rather than being subject to generation or analysis by the language grammar.

As such, this definition has certain implications in a context of oral performance. Formulaic sequences would allow both speaker and hearer to make little processing effort in their retrieval for production and understanding respectively, because they give access to "ready made frameworks on which to hang the expression of our ideas" (Becker 1975: 17). In this light, in the absence of any written form to recover some information which has been spoken, formulaic sequences are really valuable.

Formulaic expressions are often found in the recipes analysed for this study. They have been specifically labelled as statements of efficacy by Hunt (1990) and as tag phrases and efficacy phrases by Jones (1998). Alonso-Almeida and Cabrera-Abreu (2002) propose a revision of Jones's description of this type of utterances in order to clarify their classification. They dismiss the parameters of position, function and content as a means of distinguishing tag phrases and efficacy phrases because both of them appear in final position function as closing elements and are aimed at asserting the efficacy of the remedies. Instead, they advocate for an approach to the study of these formulas based on the contextual information surrounding them, so that the recipient of the text chooses a suitable interpretation that will be based on both his organisation of encyclopaedic memory and the very cognitive activity in which he is involved.

Formulas tend to appear at the end, though in some cases they can be found somewhere else such as in the title. The instance below illustrates this rather infrequent but possible placement of the formulaic sequence in question:

> (3) For þe mygreyne in þe heued þis
> medicyne is proued & for þe postem in þe
> heued & for þe dropsie in þe heued & for þe
> feuere in þe heued & for alle manere of ache
> in þe heued • Take iij peny wy3t of þe
> rote of pelleter of spayne & half a peny wy3t
> of spikenard & grynd hem to gedre

This recipe contains the formula *þis medicyne is proued* in the title. It consists of a simple sentence in the passive voice in which the agent is omitted. It is also possible to find variants of the same sequence with slight changes in the grammatical template. One of them is that in which the verbal form is in the active voice, the agent is identified with the scribe and an object pronoun is added. Thus, the sequence changes to *j haue proued hit.*

As I said earlier, formulas of the type being discussed are normally employed at the end of the remedies. The next text is an example of their occurrence in that position:

> (4) For the goute festred
> Tak auaunce archangel hayhoue beteyne
> verueyne ana & tak most of auaunce &
> boyle hem in wyn • let þe seke drynke þerof
> ferst & last & he • schal • be hol

The syntactic pattern of this formula is made up of a third person singular pronoun which refers to the patient, a verb in the future tense expressed by means of *shal/schal* and an adjective related to the lexical field of recovery from an illness. A second instead of a third person singular pronoun is used at times, as well as *wil* rather than *shal/schal*. Sometimes, these sequences may vary by the addition of phrases like *bi godes grace*.

Apart from their function as cognitive-effort-saving devices, formulas are "constructions of English which conventionally associate special pragmatic forces or effects with specific morpho-syntactic structures" (Kay 1997: 123). This definition adds a key point to the account of formulaic language and that is its pragmatic meaning in relation to the grammatical elements used to convey it. All the syntactic templates mentioned so far seem to share a unique pragmatic meaning which is to validate the efficacy of the remedy in question. The patient's recovery is said to be achieved by means of scientific procedures, as in text (3) where the writer attests the efficacy of the remedy on the grounds of someone or himself having tested it, or by means of supposed divine intervention as in those recipes where *bi godes grace* is added.

A final remark should be made as far as formulaic language is concerned. Formulas are given in the English language throughout the whole corpus. However, I have at least recorded one case in which this formula is not given in English but in Latin:

> (5) Ffor þe palsye huse bely bred in þe
> potage & non odere potage & make þat
> of seueneye & huse sauge euery day
> ffastynge & in ys drynke & annoyte
> hem with gret mostard a ȝenes a
> ffere oder sonne & drynke in ys
> potage & sanabitur

As can be observed, this recipe is all given in English except for the last word which is provided in Latin and is, precisely, a formula. This example stands as a rather interesting one as it illustrates the convergence of both oral and learned traditions. This is so because of the employment of a formula, an indicator of orality according to Ong's framework, which is written in a learned language. Therefore, it seems plausible that academic medical texts might show oral features, as well.

4.3 Redundant or copious

Repetition is favoured in oral settings because it is a way of letting hearers follow the flow of the speech event, "if only because of acoustical problems" (Ong 1982: 40). Besides, it is necessary to memorise the information that is to be conveyed through the utterances, not only from the performer's point of view, but also from the audience's, as they are ultimately the ones in charge of bequeathing this knowledge to future generations. The items that contribute to redundancy range from individual words to syntactic arrangements:

(6) For þe stone
Tak gromel • & perale • & red netle & violete
& ache & chiristones kernelles & stampe þy
kernelles in a brasyn mortere & þyn herbes
in a mortere of stoon • & þenne do hem in to
a pot of erþe & do þerto white wyn & let
hem seþe on þe fuyre • & so let hem stonde in
þe same • & 3ef þe seke drynke þerof lewe
warme & he schal be hoole

(7) For þe stone
Tak þe seed of smalache & perale &
louache • & fenel & saxfragie • & carwy
& gromel & þe rote of philependula & þe
kernelles of Chirystones ana & bete hem
to poudere & 3ef þe seke to drynke a spo-
neful at ones • in white wyn warmed
ferst & last til he be hool

Nouns that refer to plants and other ingredients, i.e. *gromel*, *perale* or *chiristones*/*Chyristones*, and cooking verbs are repeated over and over again throughout the recipes. As regards syntactic arrangements, the one that is realized in more occasions is made up of a cooking verb in the direct imperative form followed by a noun phrase or an object pronoun (*it*/*hit* or *hem*) and an adverbial expression as, for example, in (6): *stampe þy kernelles in a brasyn mortere & þyn herbes in a mortere of stoon • & þenne do hem in to a pot of erþe*.

Another dominant syntactic arrangement is made up of a verb with coercive meaning followed by a noun phrase or a third person singular object pronoun which fulfils the function of subject in the infinitive clause and an adverbial expression. This construction appears in both (6) and (7), where there are two verbs with coercive meaning: *let* and *3ef*.

4.4 Close to the human lifeworld

Closeness to the human lifeworld is highly determined by the use of deixis, which is defined by Lyons (1977: 637) as "the location and identification of persons, objects, events, processes and activities being talked about and referred to, in relation to the spatiotemporal context created and sustained by the act of utterance and the participation in it". Thus, they are linguistic items that encode the features of the context in which the speech event takes place. Deictics can be personal, spatial or temporal. Of these, I shall look at personal and spatial indexicals.

As regards personal deictics, these are the personal pronouns, all of which grammaticalise features related to person, number and gender. Specifically, personal pronouns are associated with certain speech roles (Halliday and Hasan 1976: 45) and so, they define the participants engaged in a communicative situation. There are two main roles and these are addresser and addressee. Other roles can be found, as well, and they include the rest of entities relevant to the speech situation other than addresser and addressee.

The recipes analysed contain a large number of personal deictics; the most frequently employed is the third person singular pronoun *he/hee*, whose referent tends to be the person that needs to be cured and whose syntactic function is that of the subject. The feminine counterpart of this pronoun, *sche*, only appears in very rare occasions, just in those cases where the text deals with gynaecological diseases.

The next pronoun in the frequency scale is *þou*, which refers to the person being addressed and, in general, those who might find the remedies presented in these texts useful in order to help someone recover from an illness. Linguistic evidence signals an English-speaking person as addressee. The recipient of the recipes does not need to have much knowledge in herbal and medicinal matters since the writer leaves explanations and descriptions here and there to guide his folk audience. The recipe shown next illustrates this point:

> (8) And also for to make a
> schorned hed hool Take sewet of a dere
> & an herbe þat groweth on þe grounde þat
> men calleþ ffeld draf & þat is lyk hony
> combes tak þat & sewet & fferye hym to
> gedre • & strayne hit þour3 • a clod & tak þat
> leukere & ley on þe hed with a federe & þat
> schal do away þe schornes of þe heued •
> & afterward ley on of þe lewkere before
> jseid, other feyr hony & þat schal make
> hit hool • bi • godes grace • certayn, for j haue
> proued hit

The first sentence contained in the body proper of the recipe tells the addressee about two ingredients that are needed to make it, namely, *sewet of a dere* and another herb for whose identification he provides a description by means of three relative clauses introduced by *þat*. These chunks of information might be considered to be a kind of help for potential users who lack knowledge about certain specific herbs. It should be noted that although the vast majority of the texts are written in English, there are some fragments written in Latin. This restricts the audience of those passages to users with some knowledge of it.

In relation to the first person singular pronoun, I have recorded only two occurrences of it, both of which are worth commenting on. The first of them appears in example (8), too, particularly in the last sentence of the recipe which in this case corresponds to the efficacy statement. As was previously stated, it contains a future tense indicated by *schal*, as well as the phrase *bi godes grace* and an adverbial clause of reason introduced by *for* whose subject is the pronoun *j*.

The use of this pronoun stands as an obvious case of self mention thanks to which the author's presence is made explicit. It is an interactional resource that can be found in metadiscourse, which, in turn, is to be understood as a "cover term for the self-reflective expressions used to negotiate interactional meanings in a text, assisting the writer (or speaker) to express a view point and engage with readers as members of a particular community" (Hyland 2005: 37). Therefore, the use of this pronoun can be considered to be a metacomment whose pragmatic function is that of reinforcing the validity of the remedy, already stated by the presence of the statement of efficacy into which the adverbial clause is embedded. Furthermore, the addresser also makes use of the item *certayn* which is a booster. Its use results in a metacomment that emphasises the speaker's certainty in relation to the propositional content expressed, that is to say, the state of affairs described (the patient's becoming healthy) will definitely happen.

The other instance of self mention appears in the recipe below:

> (9) Fore þe jawnes a good medicyne proued
> Tak a red cok þat is ry3t feres & good fe3ter
> & putte him vnder a bosschel or vnder a lepe
> al a ny3t & a morwe tak þe donge þat he haþ
> made & do it in to a vessel & tak as muche
> juory & as muche saffran & do þerto & of ech
> j liche muche & grynde alle þese iij þyngis
> to gedre smale & þenne strayne hem þur3 a
> cloþ & 3ef þe seke drynke þerof ix dayes but
> lat no3t þe seke wite what it is for þanne
> j suppose he wolde nou3t drynke þerof but 3ef
> hym drynke þerof & he schal be hool vp wa
> rantise þur3 godes grace

j suppose is another metacomment in which the producer of the text makes use not only of self mention, but also of hedging. This is another interactional resource which allows him to present a new chunk of information as an opinion rather than as a fact. Consequently, the propositional content that is attached to the whole utterance in which *j suppose* appears is uncertain: the writer imagines that if the physician let the patient know what the remedy is for, he would not drink the beverage. Since the writer is not completely confident in this possibility, he employs a hedge which ultimately secures his position.

The feature of closeness to the human lifeworld as indicated by the employment of personal deictics overlaps with the status of orally based expression as empathetic and participatory rather than objectively distanced. This is so because personal pronouns, above all, *þou*, mark engagement between the participants in discourse: these pronouns are the means by which the producer of the recipes addresses their recipient directly creating a bond between them in two senses, that is, authoritative and affective. The occurrence of a second person pronoun evidences the addresser's position in relation to the addressee. The position is one of acknowledgement of the addressees's being necessary participants in the communicative situation.

In relation to spatial deictics, these are considered to be a special kind of deictic terms as they encode the speaker's ego (Wu 2004: 33): any spatial relation between utterance and context is indicated from the speaker's perspective. That is the case of demonstratives and adverbs, all of which are key items in order that one might have points of reference in space. For instance, when hearing *this* or *that*, the addressee would search for the referent of these demonstratives in the speaker's vicinity and farther away the speaker's vicinity, respectively. This recipe for dropsy contains an example of spatial deixis:

(10) For þe
dropsie in þe wombe & in þe feet & for
costifnesse & for glete almite þe stomak
& for wormes in þe wombe & for a ladde
stomak Tak iij peny wi3te of stamem
& ij peny wi3t of rubarbe & vj peny
wi3t of Cene & iij peny wi3t of spike
narde & iij peny wi3t of floure of Ca •
nel • & half an vnce sugre of Cipres
& do hem to gedre • & bray hem in a bra-
syn mortere alto powder • & let þe seke
vse þis medicyne fastynge euery day
a good sponeful iij tymes in þe day &
he schal be hool

This recipe contains the spatial deictic *þis* in the application section where the writer tells the audience how to give the medicine to the sick. It refers to something which has already been mentioned and which stands in its vicinity, otherwise the writer would have used *þat*. It particularly refers to the whole mixture of ingredients that have been made into powder and have been treated in a very specific way to finally achieve a compound medicine. The use of *þis* as a spatial indicator is completely contextually dependent as its referent varies with each speech event.

The adverb *here* functions as a spatial indicator, as well. It is used in several recipes as in another one for dropsy:

(11) Here is
þe hei3est medicyne for þe dropsie
Summa medicina per ydropicer aqua distilla
ta • de floribus sambuci videlicet of þe flou
rys of ellerne temperatur per se per <...> ydropsie
discum coclear plenum mane & sero

Here is a circumstantial demonstrative in which the addresser "identifies the referent locating it on a scale of proximity" (Halliday and Hasan 1976: 57) and, in particular, it refers to something that is situated nearby. In this case and being closely related to the idea of physical immediacy, *here* means in this precise place and is used cataphorically since its referent is situated immediately after it, as opposed to the ways in which deictics are used in the examples above (anaphorical reference). *Here* also fulfils a textual function as it is used to introduce another remedy by placing it in the title.

The property of spatial deictics as achieving full meaning when the context is taken into account turns them into situational rather than abstract items, in which case there is overlap again between two traits of oral expression. In both closeness to the human lifeworld and situational rather than abstract features, it is suggested that even though indexicals have an abstract meaning attached to them, it is not until the specific circumstances in which a speech event takes place that the propositional contents conveyed by them are completely understood.

5. Conclusion

In this paper, I have explored the presence of orality in a corpus of Middle English medical recipes which have been excerpted from G.U.L. MS Hunter 185 and belong to the research project *Las Palmas Corpus of Early English Recipes*. The reason why I have chosen a remedy book for this study is that it has been traditionally assumed that both their origin and audience were popular and so, they might be an appealing starting point for this type of research in medical texts. In order to do so, I have followed Ong's proposal of features of orality, especially four of them: (i) additive rather than subordinative, (ii) aggregative rather than analytic, (iii) redundant or copious and (iv) close to the human lifeworld.

This first approximation has yielded some conclusions, the most relevant of them being that the recipes scrutinised have, in fact, preserved some features of orally based expression. They have proved to be syntactically constructed by simple grammatical patterns, mainly by an observable frequent use of coordination instead of sub-ordination. Hypotactic arrangements, however, are not absent and their meaning tends to be adverbial in relation to the main clause.

The presence of orality has also been attested in the corpus by the occurrence of formulaic expressions whose pragmatic meaning was to assert the efficacy of the recipe when treating patients. The grammatical templates which convey this meaning can be varied such as (i) noun phrase + *is proved*, (ii) first person singular pronoun + *have proved* + object pronoun, and (iii) third person singular pronoun + *schal be* + hool.

Repetition of lexical items and syntactic constructions are typical of oral expression, too, as they help the audience follow the flow of speech and memorisation. Names of plants and ingredients used in medical preparations as well as cooking verbs are repeated in the texts analysed. They are not only indicators of lexical and grammatical cohesion in the individual recipes and the whole remedy book at the same time, but also of the redundancy associated with the oral mode.

As for closeness to the human lifeworld, it has been observed that this trait of orality is linguistically manifested by deictics. Personal pronouns, above all, first person singular *j* and second person *þou*, signal the presence of addresser and addressee as participants in discourse. The author's stance might be said to be both authoritative and empathetic. The former can be identified thanks to the explicit use of *j*, though in rare occasions, and the latter is associated with *þou* which he employs to acknowledge the addresees's need of guidance in the speech event.

Another conclusion derived from this study is that, apparently, there seems to be some sort of connection between supposed popular medical texts and oral tradition. Nevertheless, the nature of that connection cannot be fully specified until the corpus of study is enlarged so as to get quantitative results which support this preliminary conclusion. Future research should also include academic medical writings to carry out contrastive analyses as a way of characterising medieval texts in general in terms of their having retained or not features of orality.

References

Alonso-Almeida, Francisco 2000. *Edition and Study of a Late Medieval English Medical Receptarium: G.U.L. MS Hunter 185 (T.8.17)*. University of Las Palmas de Gran Canaria: Las Palmas de Gran Canaria.

——— 2008. The Pragmatics of And-Conjunctives in Middle English Medical Recipes: A Relevance Theory Description. *Journal of Historical Pragmatics*. 9/2, 171-199.

Alonso-Almeida, Francisco / Carroll, Ruth 2004. A New Proposal for the Classification of Middle English Medical Texts. In Rodríguez-Álvarez, Alicia / Alonso-Almeida, Francisco (eds.) *Voices on the Past. Studies in Old and Middle English Language and Literature*. A Coruña, Spain: Netbiblo, 21-33.

Becker, Joseph 1975. The Phrasal Lexicon. In Shank, Roger / Nash-Webber, Bonnie (eds.) *Theoretical Issues in Natural Language Processing*. Cambridge, Massachusetts: Bolt Beranek & Newman, 60-63.

Carroll, Ruth 2003. Recipes for Laces: An Example of a Middle English Discourse Colony. In Hiltunen, Risto / Skaffari, Janne (eds.) *Discourse Perspectives on English: Medieval to Modern*. Amsterdam and Philadelphia: John Benjamins, 137-165.

Chafe, Wallace 1982. Integration and Involvement in Speaking, Writing, and Oral Literature. In Tannen, Deborah (ed.) *Spoken and Written Language: Exploring Orality and Literacy*. Norrwood, New Jersey: Ablex, 35-53.

Halliday, Michael / Hasan, Ruqaiya 1976. *Cohesion in English*. London and New York: Longman.

Havelock, Eric 1963. *Preface to Plato*. Cambridge, Massachusetts: Harvard University Press.

——— 1977. The Preliteracy of the Greeks. *New Literary History*. 8/3, 369-391.

——— 1991. The Oral-Literate Equation: A formula for the Modern Mind. In Olson, David / Torrance, Nancy (eds.) *Literacy and Orality*. Cambridge: Cambridge University Press, 11-27.

Hyland, Ken 2005. *Metadiscourse. Exploring Interaction in Writing*. London and New York: Continuum.

Hunt, Tony 1990. *Popular Medicine in Thirteenth-Century England*. Cambridge: D.S. Brewer.

Jones, Claire 1998. Formula and Formulation. “Efficacy Phrases” in Medieval English Medical Manuscripts. *Neuphilologische Mitteilungen*. 99/2, 199-209.

Kay, Paul 1997. *Words and the Grammar of Context*. Stanford, California: Center for the Study of Language and Information Publications.

Kroll, Barbara 1977.Combining Ideas in Written and Spoken English: A Look at Subordination and Coordination. *Southern California Occasional Papers in Linguistics*. 5, 69-108.

Lord, Albert 1960. *The Singer of Tales. Harvard Studies in Comparative Literature, 24*. Cambridge, Massachusetts: Harvard University Press.

Lyons, John 1977. *Semantics (Volumes 1 and 2)*. Cambridge: Cambridge University Press.

Ong, Walter 1982. *Orality and Literacy. The Technologizing of the Word.* London: Methuen.

Overstreet, Maryann /Yule, George 2001. Formulaic Disclaimers. *Journal of Pragmatics*. 3, 45-60.

Parry, Milman 1971. *The Making of Homeric Verse: The Collected Papers of Milman Parry*. Oxford: Oxford University Press.

Taavitsainen, Irma / Pahta, Päivi / Mäkinen, Martti 2005. *Middle English Medical Texts*. Amsterdam and Philadelphia: John Benjamins.

Tannen, Deborah 1982. The Oral/Literate Continuum in Discourse. In Tannen, Deborah (ed.) *Spoken and Written Language: Exploring Orality and Literacy*. Norwood, New Jersey: Ablex, 1-16.

——— 1986. Spoken and Written Narrative in English and Greek. In Tannen, Deborah (ed.) *Coherence in Spoken and Written Discourse*. Norwood, New Jersey: Ablex.

Voigts, Linda 1982. Editing Middle English Medical Texts: Needs and Issues. In Levere, Trevor *Editing Texts in the History of Science and Medicine*. New York: Garland, 39-68.

Voigts, Linda / Kurtz, Patricia 2000. *Scientific and Medical Writings in Old and Middle English: An Electronic Reference*. Ann Arbor: University of Michigan Press.

Wu, Yi'an 2004. *Spatial Demonstratives in English and Chinese. Text and Cognition*. Amsterdam: John Benjamins.

IRMA TAAVITSAINEN

Early English Scientific Writing: New Corpora, New Approaches[1]

1. Introduction

What philology has to offer to corpus linguistics is expertise in corpus compilation, in terms of both text selection and detailed attention to context in the interpretation of texts, where meanings have to be negotiated at several levels. In return, corpus linguistics can raise textual studies to a new level. On the basis of qualitative studies, we can make claims about the practices of earlier discourse communities based on individual instances of real language use, but it is through quantitative analysis of large databases of authentic language that these claims can be validated and subjective judgements brought to a more objective level.

Recent years have seen new interfaces emerge between traditional philology and corpus linguistics, and a fruitful exchange of ideas is going on. Advances in the development of electronic tools are the underlying factor in this new phase. These new tools and new approaches make it possible for researchers to approach Middle English medical and scientific writing with new accuracy.

1 Research for this paper was funded by the Academy of Finland (project number 1118478), and carried out at the Research Unit for Variation,Contacts and Change in English at the Department of English, University of Helsinki.

2. Approach and aim of the article

My approach is one of historical pragmatics and, more specifically, of historical discourse analysis, with a focus on diachronic developments. Historical pragmatics is an empirical branch of study which focuses on communication and the contextual assessment of language use in historical periods. The aim of this article is to contribute to the above-mentioned exchange of ideas, and to show in detail how corpus-linguistic methods can be combined with qualitative contextual assessments. The focus of my presentation is on late medieval medical writing, but I shall also transgress the boundary with early modern texts. I shall discuss new developments in corpus compilation, and how the new material collected using these developments can be applied to answer new research questions that deal with the diachronic development of genres and discourse forms and, even more profoundly, with the changing thought-styles of medical writing. My aim is to present some new insights and demonstrate how a more diversified picture of the developments in the medical register has emerged, with an emphasis on the communicative function of genres in the sociohistorical framework of the period. The investigation focuses on the changes in genre between the late medieval and the early modern period. The reasons behind these changes are often changes in background factors. By relating the results of linguistic searches and queries to contextual assessments and the parameters of the communicative purpose and context of production, we can gain new insights into changing thought-styles in the history of English.

3. Methodology

3.1. Combining corpus linguistics with philology

The advent of electronic text collections has made the retrieval of data for linguistic studies speedy and effective. With the appearance of corpora of texts, the whole research paradigm of linguistics has changed: the focus of linguistic analysis has shifted from individual examples of structures that in reality could be rare to actual language in use. The arm-chair linguist has given way to the empiricist; corpus linguists do not believe claims about language use without empirical evidence, preferably with statistical backing. Furthermore, large linguistic databases lend new accuracy to linguistic studies if we apply them with an analytical grid which is specific enough to catch the differences in the relevant features. It is also possible to find answers to new research questions using these approaches.

Diachronic discourse analysis is concerned with the functions of discourse features and the structures of larger units of text in a diachronic perspective. It is a fairly new branch of study with a fresh outlook: discoursal changes are seen to be on a par with phonological, morphological, syntactic, semantic and pragmatic change. In both this and other branches of historical linguistics, it is possible to achieve an overall view of diachronic developments by combining synchronic descriptions of various periods. Corpus-linguistic methods offer us a means to collect well-documented empirical data about the distribution of linguistic and stylistic features.

It has been an accepted tenet for about twenty years that it is the co-occurrence patterns of linguistic features that define the style of a text. Several features work together to create the style. A groundbreaking corpus-linguistic study with refined statistical methods established patterns of variation in multidimensional space, the most important being "Involved versus Informational Production", "Narrative versus Non-narrative Discourse" and "Situation-dependent versus Elaborated Reference". This study also indicated where genres of spoken and written language would fall along these continua (Biber

1988).[2] The study has inspired a host of others and, in general, corpus stylistics is an exciting area of new developments. New corpora with new and more comprehensive materials can give us more detailed information about developments within specific areas of language use, such as the medical and scientific register of writing. Another innovation brought by corpus linguistics is the ability to combine the results offered by searches on electronic corpora with more qualitative studies and information about background factors.

Traditional philology with its emphasis on context has a great deal to offer to our corpus studies on early scientific and medical writing. *Context* here denotes a very rich concept which includes ideological, historical, material and textual contexts. We as empirical linguists are interested in language use in context, and methods and principles derived from careful textual studies should complement our corpus linguistic studies. Central to the philological approach is the assumption of linguistic skill, expertise and profound knowledge of the language form, sometimes called the "native speaker competence" of Old English and Middle English. A detailed knowledge of the language form under study, along with knowledge of its sociohistorical background and historical developments, is a prerequisite for successful analysis in philological studies, but it is just as important in corpus-linguistic research on older stages of the language. Texts are produced in historical contexts to meet communicative needs. Language-external facts must be recovered in order to be able to interpret how people in the past negotiated meaning in these texts, and we need a contextual approach to ensure the integrity of our studies (Taavitsainen and Fitzmaurice 2007).

---

2 The study was based on 67 linguistic features in 450 texts of various genres. A study applying the same method to scientific writing in a historical perspective was conducted by Atkinson (1999), who chose the Royal Society period as his starting point.

3.2. Applying corpus-linguistic methods to medieval medical writing

At this point, I would like to return to corpora and corpus-linguistic methods and relate them to the study of late medieval vernacular genres. The point of departure is the research question, what we want to discover. The aim of our pilot studies was to establish features that define the scholastic thought-style, and we relied on descriptions of medieval scholastic science by historians of science. The source of knowledge for these medieval scientists was "that someone said so", i.e. the quotative, "hearsay" was the mode of knowing, and there was a firm reliance on axioms derived from ancient authorities. For our pilot studies on scholastic style, we chose linguistic features that would be the textual representation of these features. In order to create a list of the relevant search items, we consulted the wordlists made from our corpus texts by an earlier version of the software package MEMT Presenter. Word lists are indispensable at this stage for ensuring the inclusion of all manifestations of the items in question. Our overall approach was "top down", but at this stage we resorted to the "bottom-up" method as well. Our hypothesis was that we could probe deeper into the style of writing by assessing the following linguistic features:

> Reliance on authorities > references to them > names of authorities in the word lists > corpus searches
>
> Quotative evidence > speech act verbs of saying > word lists of texts > corpus searches
>
> Deontic modality > prescriptive phrases > it is + V (semantic classes) > corpus searches

Our pilot studies indicated that there are systematic differences in the styles of writing, and in the text-type features of the genres. The results of our corpus searches revealed the layered nature of medical learning, and we were able to establish a hierarchy of authorities in vernacular texts. By investigating the background information and combining it with our results, we were able to draw conclusions about the correlations between the level of writing and the references made. Galen and Hippocrates are frequently mentioned in learned and specialised treatises; next in frequency in learned texts come Arab authors, such as Avicenna, Rhases, Haly Abbas and Averroes, and medieval Latin authors; general references to doctors, leeches, physicians and masters prevail in more popular registers of writing. The searches for prescriptive phrases indicated similar differences, with cognitive verbs prevailing in the learned registers and concrete action verbs in the texts for more heterogeneous audiences (Taavitsainen and Pahta 1998: 169, 1997).

I have taken the bottom-up method as my point of departure in two studies, combining philological assessment, i.e. careful qualitative reading, with corpus searches. The first dealt with modality (Taavitsainen 2001a). Modality is a demanding and elusive category to study, and thus the bottom-up method seemed particularly appropriate for understanding the ways in which modality was expressed in late medieval medical texts. As a first step, I made a small stratified test corpus for manual assessment. It included texts from various genres and traditions, representing the full continuum of medical writing in Late Middle English. This subcorpus was then studied carefully using the "philological" method to find out what means were used to express modality. I then performed searches for the features I had chosen in the electronic corpus in order to provide a wider perspective.[3] More recently, I conducted a study on metatextual features using the same bottom-up method (Taavitsainen 2006). This led to a categorisation that differed from other assessments made on recent materials (Hyland 2005). I shall continue my studies on the development of metatext within early modern medical texts in a forthcoming study.

In the future, some interesting results and new insights can be expected from the application of new corpus-linguistic tools and software; for instance, the keyword function of WordSmith 4 provides a new tool for detecting discourse features in texts. The difficulty in applying this method lies in the inherent nature of Middle English, with its abundance of spelling variations. The prerequisite for the

3 cf. Kohnen forthcoming 2008.

application of the keyword function is consistency of spelling in both the texts under scrutiny and the relevant test corpus. Creating a solid basis in this way takes a great deal of background work on both Middle English and early modern texts. The materials should be normalised to give reliable results and some studies have already been made that address this problem (see, for example, Ratia forthcoming, Kohnen 2009).

4. Middle English Medical Writing

4.1. Changing conceptions of science and the medieval world view

Science has been defined in various ways, but basically it deals with the consideration of our world. Science includes knowledge of the order of the world, natural phenomena, and the laws that govern our existence. This fundamental aspect of culture is reflected in the extant written documents from the medieval and early modern periods, both literary (e.g. Chaucer's works) and non-literary. The position of medicine among the sciences is a special one because of the problem of the relationship between theory and practice in this area. This problem is an old one, and occupied learned writers for centuries throughout the Middle Ages and beyond. Medicine was both a learned university discipline and an occupation involving technical skills. Medical education comprised general theories and specialist doctrines, but also practical applications through apprenticeship (Crisciani 2000: 75-76). The borders between science, pseudoscience and the occult were drawn in a different way in the late medieval period than they are today. In that period, astronomy/astrology was a branch of progressive mainstream science, a cornerstone of Western medicine,[4]

4 The difference between science and pseudoscience has been defined by Crombie (1995) in terms of change: when science ceases to change, it ceases to be science. His view applies to astrological and alchemical texts in particular. These cannot be considered a science by modern criteria, as the

and belief in celestial influences was pervading. Medical astrology provided the means of defining the correct times for various actions, such as taking medicine or undergoing operations, as well as for non-medical issues. Such advice was important both for the learned and for those who had to resort to simplified versions of astronomy based on observations of the skies without computational devices like the astrolabe or the volvelle, a device used by medical practitioners for timing their actions. According to the medieval world view, man was the centrepiece of the universe, the microcosm, at the mercy of the influences of the macrocosm of the universe. An extensive system of correspondences formed a coherent body of theories and practical applications: planetary control extended over the parts of the body as well as over all other things in the world. Knowledge of medical doctrines was disseminated through various layers and genres of writing in late medieval society, reaching the widest possible audiences in almanac literature (see below). The adherence to astrology became even stronger in the Renaissance, and the same doctrines, albeit in a stilted form, linger on as a pseudoscience even today, with similar applications.

4.2. The emergence of scientific and medical writing in English

The discourse world of early medieval science was a monolingual one which spoke and wrote only Latin.[5] The theoretical foundations of

---

basis for them has been proved false. Alchemy is perhaps the extreme case in the changing paradigm of science: in Chaucer's time it was a core science, but three centuries later, in 1704, it was ironically described as an art which begins with lying, is continued with toil and labour, and at last ends in beggary.

5 Latin retained its position as the language of the majority of medical publications until the end of the seventeenth century in England (Webster 1979), and served as a *lingua franca* even much later. In England, Anglo-Norman texts started to appear somewhat earlier than English texts. MEMT includes an appendix from c. 1330, a trilingual wordlist, after which there is a gap of c. 150 years before the appearance of the next vernacular texts (see MEMT Introduction 2005).

medicine derived from antiquity and were modified by Arab scholars, who systematised the links between medicine and philosophy. Italy became the focal point for the formation of the traditions and conventions of Western science and medicine (Siraisi 2001). The twelfth century is considered the starting point for a long medical Renaissance which evolved over four hundred years, eventually resolving a broad range of issues and defining a new science of medicine, which grew out of a range of equally new areas of scholarly investigation (McVaugh and Siraisi 1990, 9). These new ideas were expressed in Latin texts in learned centres like Salerno in the twelfth century. Vernacular versions of texts started to appear somewhat later, in the thirteenth century in France and the fourteenth in Spain, Portugal, Germany and England (see Crossgrove 1998, and Taavitsainen and Pahta 2004). Genres of writing were transferred into the vernaculars following Latin models. Glimpses of the underlying motivations of the process can be found in the metatextual genres of prologues and introductions. On the one hand, translations were inspired by nationalist feeling and the desire to enhance the status of the vernacular language; on the other hand, they might also have charitable motivations. Prefatory materials often mention that the leading motivation was to make medical advice available to "the poor", to improve their condition by giving them access to useful knowledge about medicines and healing practices (Wogan-Browne 1999, Getz 1990). All levels of medical texts were involved in the process: there were translations and adaptations of learned specialised treatises and surgeries, compendia with theoretical knowledge and practical applications, and miscellaneous collections with recipes and remedies, rules of health, prognostications and astrological tracts. There is still a great deal that we do not know about the extent of this process. The underlying manuscript reality comprises a great deal of unexplored material which represents genuine possibilities for new discoveries.

4.3. New tools for studying the vernacularisation of medical and scientific writing

The opportunities for making new discoveries and achieving a reliable overview of the sources and connections of texts have improved greatly as new research tools have been made available to scholars within the last decade. This is one of the reasons why we can, without exaggeration, say that studies on medieval scientific writing rest on a new foundation and have entered a different phase. The database *Scientific and Medical Writings in Old and Middle English: An Electronic Reference*, by Voigts and Kurtz (eVK, 1998), and George Keiser's volume *A Manual of the Writings in Middle English 1050-1500: Works of Science and Information* (1998) are invaluable for charting the extent of English medical writing and give us a good overview of the situation. Information about the underlying manuscript reality will become easily accessible with the publication of the *Electronic Thorndike Kibre Catalogue* (eTK, forthcoming). A different type of tool, the electronic corpus *Middle English Medical Texts* (MEMT), was released for public use in December 2005, giving access to a large collection of Middle English medical texts with a full coverage of all traditions and genres. It is the first part of the *Corpus of Early English Medical Writing* (CEEMW 1375-1800).[6] I shall discuss the decisions underlying its compilation, and demonstrate using empirical examples how I have applied corpus-linguistic methods to pragmatic and discoursal research questions, which have traditionally been outside the scope of corpus studies.[7]

---

6 MEMT is the first of a series of three corpora. *Early Modern English Medical Texts* (EMEMT) was released in 2009, and work on *Late Modern English Medical Texts* is also underway. The current project members are: Irma Taavitsainen and Päivi Pahta (senior scholars); Turo Hiltunen, Ville Marttila, Maura Ratia, Carla Suhr, Jukka Tyrkkö, Alpo Honkapohja (doctoral students); Anu Lehto (research assistant); Martti Mäkinen (post doctoral researcher, Stavanger).

7 E.g. pragmatic research questions have been brought to the attention of scholars as the central themes of important conferences: IPrA 2007 had Pragmatics and Corpus Linguistics as its theme, and the ICAME Conference

4.4. The state of the art of research on medieval scientific and medical writing

There seems to be a paradigm shift going on in the humanities: team work is replacing the individual scholar's lonely toil in the ivory tower. In the field of medical and scientific writing, the pioneer team was the Scientific Thought-Styles (STS) project, launched in Helsinki some fifteen years ago with the aim of charting scientific thought-styles in medical writing in a long diachronic perspective. Our research has shown that the linguistic features typical of scholasticism and empiricism are very different from one another, reflecting the underlying methodology and philosophy of science (see below). A pioneer study on medical and scientific writing in the *Helsinki Corpus* revealed a development from a more detached to a more involved style of writing between the fifteenth and the late seventeenth centuries. It also showed that there was a great deal that we did not know yet in this field (Taavitsainen 1994). The study gave us an incentive to develop an ambitious goal, which is still at the heart of our research agenda: to investigate, within a sociohistorical framework, the development of medical writing from scholasticism to empiricism and beyond (Taavitsainen and Pahta 1995). Different periods are traditionally connected with different styles of thinking and making decisions, and it is clear that the question of what continued and what changed is central to the historical analysis of scientific thought-styles (Crombie 1994: 6). The next step is to relate the types and causes of variation to the underlying ideologies, scientific methodologies, traditions of writing, education level of the authors, and levels and scope of the audience (see below). We are still following the same pathway, but it has become broader with the inclusion of new research questions on discoursal and pragmatic topics (see below), and there are more of us to do the research. Furthermore, our project is not alone in its efforts, as there is a great deal of interest in the history of scientific writing. Researchers in Spain are active in the field. There are at least two projects that focus

---

(14-18 May 2008) was devoted to the theme of Corpora: Pragmatics and Discourse.

on the late medieval period. An international research project with participants from Glasgow, Malaga and Oviedo is aiming to produce electronic editions of the late medieval scientific manuscripts housed in the Hunterian collection at Glasgow University Library and to build an annotated corpus of Middle English *Fachprosa* based on these editions. Scholars in Gran Canaria are actively researching medieval recipes, compiling a corpus and studying the linguistic features of these texts (see below). In addition, individual scholars are pursuing studies on medieval medical writing. The Research Group for Multidimensional Corpus-based Studies in English (MUSTE) at the University of A Coruña focuses on the diachronic development of the scientific register from the seventeenth century onwards. This group is also actively compiling corpora of non-medical scientific writing. In Italy, scientific writing is an important research topic, although Italian research focuses on more recent phases. There is a great deal of ongoing editorial activity and manuscript studies of medical texts in America and Britain (see below). This is a good situation: we can exchange ideas and collaborate towards a common goal of rewriting the history of scientific and medical English more accurately, taking different genres, fields of study, and traditions into account.

4.5. Textual basis, text selection, and corpus design in MEMT

MEMT is mainly compiled of edited medical treatises, but there are some new transcriptions of texts as well. We collected all edited texts that we knew of.[8] We digitised them manually or scanned them for MEMT, extracts of c. 10,000 words of longer texts and shorter texts in whole. In addition, we made some new transcripts of important

8 A great deal of research went into this phase, first in charting the field and then in acquiring the copyrights. The new tools mentioned above were invaluable for providing information about texts and editions in the field. Transferring the texts into an electronic corpus was a huge task, and we had several helpers and student assistants working for us over the years. We are grateful to the Academy of Finland, the University of Helsinki, and the Research Unit for Variation and Change in English (Varieng) for making this work possible.

treatises, as it turned out that some early editions had been made for the needs of groups such as historians and were thus not suited to linguistic studies (see MEMT Introduction), or because we felt that there was a real gap that could be filled in this way. MEMT provides a great deal of fresh material that was earlier unavailable to scholars, such as several previously unpublished theses.

We divided the material according to the traditions underpinning the writing, and defined it as a continuing series of texts which built on one another, with a great deal of intertextuality. The main branches of the corpus tree in MEMT are defined according to the traditions of "Surgical texts", "Specialized texts", and "Remedies and *materia medica*", but genres are not indicated in MEMT, as a great deal of research is needed before such a categorisation can be devised (see below). Texts in the first and second categories are learned, and have their origins mostly in the academic tradition. The transmission of texts is complicated, however, and we cannot make the distinction between academic and popular texts by relying on modern notions.[9] The third main group includes remedy books, recipes, prognostications, and *materia medica*, represented in the corpus mainly by herbals, but the source material for such texts also included learned materials, and there is evidence that the users of these books included medical practitioners at the top of the hierarchy alongside more heterogeneous audiences. Verse in this period was not as sophisticated as prose. We have included practical poems on medical topics in the corpus to show the scope of medical writing, but these are recorded as a separate group.[10] Our stopping point was in autumn 2005, and the collection of texts that had been edited before that time is as comprehensive as possible.

9 'Popular' is here defined as describing a particular way of using texts as cultural products; see Taavitsainen 2005 and forthcoming.

10 See MEMT Introduction for details.

4.6. Editions and the textual world of medieval scientific and medical writing

Reliable, well-documented editions are invaluable for our research on earlier stages of language, as they can provide us with the information that we need to make sound judgements. In recent years, the philosophy of editions has changed, and electronic tools have opened up new possibilities. Researchers are becoming aware of the idea of multiple readings, and a more open attitude to texts is gaining ground; fuzzy sentence boundaries, different scopes of modification, indeterminacy of discourse features like direct and indirect speech and narration, and multiple voices in texts are among the characteristics of medieval texts that cause major problems of interpretation and require more study from the pragmatic point of view (see Taavitsainen and Fitzmaurice 2007; Lass 2004, Robinson 2005).

When interest in philology first flourished in the latter half of the nineteenth century, the texts selected for editing were mostly literary or religious, but some of the more important surgical texts were also made available to scholars in the first boom, and the nineteenth-century editions of Arderne's *Fistula in Ano, Haemorrhoids, and Clysters* by Power and Lanfrank's *Science of Cirurgie* by Fleischhacker (Early English Text Society O.S. 139 and O.S. 102, respectively; extracts of both included in MEMT) are still the only ones available for these texts. Some remedies were also edited because of their curiosity value, but in general, the shift of interest towards scientific, utilitarian and everyday texts only took place about twenty years ago, and the trend has been increasing since then. New electronic editions build on new principles and make use of advanced technology.[11] In addition, textual studies on medieval manuscripts add to our knowledge of textual affinities, scribal practices, and medieval book production. A recent achievement in this

11 E.g. one project group of three doctoral students working on their dissertations in Varieng is called "Digital Editions for Corpus Linguistics: Representing manuscript reality in electronic corpora". This groups is working to apply XML coding to early texts and make use of the new possibilities for making editions of medieval and early modern texts.

field is the two-volume study of the texts in MS Trinity College, Cambridge R. 14 (Tavormina (ed.) 2006), which includes editions of these texts.[12]

4.7. Medical writings and their audience

The question of the target audience for vernacular texts is an important one, as, according to the definition of *genre*, the development of a genre depends on the audience: genres are created to meet the needs of discourse communities (Swales 1990, see below).[13] In the late medieval period, the medical profession consisted of heterogeneous groups of practitioners with different educational backgrounds. Physicians, surgeons, and perhaps even some apothecaries, had a university education. They were the readership for the highest-register genres of Latin medical texts and their vernacular counterparts (see below). This literature included the genres of commentaries and *compilationes*, encyclopaedic treatises, *questiones*, pedagogical dialogues, and *consilia* and *practica*, which were basically case reports. In addition, medical practitioners at the highest level are known to have owned and used books belonging to the remedy-book tradition.[14] Barber-surgeons established guilds and formalized their training in the late Middle Ages. We have first-hand knowledge of genres in use in the shape of the Guild Book of the Barber-Surgeons of York. This document contains treatises on the basic doctrines and their applications, including several types of fairly simple prognostications. Midwives, meanwhile, were women with practical training and experience. Some texts from their field have been preserved, and some of these were explicitly written for women

12 Two of the texts in the above manuscript are included in MEMT: *De spermate* (Pahta 1998) and *De humana natura* (Carrillo Linares 1993). The others were not available to us in 2005.

13 The term was originally coined for applied linguistics, but it can well be extended to medieval medical literature and to groups of people connected by texts, either as a community or individually. For a discussion of medieval medical discourse communities, see Jones 2004, 23-26.

14 The classic example is John Argentine, King's College (Rhodes 1956).

(see Pahta and Taavitsainen 2004, 17, Barratt 2001). Itinerant specialists like bonesetters and oculists were mainly a feature of the rapidly growing towns; in addition, there were herbalists, wisewomen, and others with heterogeneous backgrounds. These groups were largely illiterate and learned their trade through practice. Literacy was restricted mostly to the elite group, but in the fifteenth century, the rising upper middle classes were becoming increasingly literate and eager to learn and improve their knowledge of useful matters like health (Siraisi 1990). They formed a new type of audience for the handbook and health guide literature that started to flourish in the fifteenth century. This trend became even more pronounced in the early modern period.

5. Medical discourse, traditions and genres of writing: From late medieval to early modern

The term *medical discourse* in this context is a collective term referring to the communicative practices of the medical profession, including their traditions and genres of writing (and also spoken genres, but these are beyond the scope of historical assessments of early periods). Traditions of writing, defined as a continuing series of texts building on one another, form the underlying main current, and each comprises several genres. Genres fulfil communicative functions in society, and they are important operational tools and constitute dynamic systems which undergo change and variation. Over the course of time, sociocultural needs change and genres change accordingly: old genres become adapted to new functions, new genres are created, and genres that have lost their function cease to exist. Thus *genre* refers to a grouping of texts that is based on function and text-external evidence, whereas *text type* is a more technical term used of groupings made according to the text-internal features. I have found this distinction useful in my studies on early genres of medical writing (see Taavitsainen 2001a).

The commonly used date for the transition point between the medieval and early modern periods, 1500, is more conventional than real, as the medieval textual world continues without much change in the first half of the sixteenth century. New ideas about the world appeared at the dawn of the sixteenth century and were adopted and developed at different times in different branches of science. New knowledge diffused slowly through the layers of scientific writing, existing within the frame of the older knowledge and mixing with it. The basis of medical treatment started to change with increasing knowledge of anatomy and physiology, and in the latter half of the sixteenth century there was a growing lack of faith in the wisdom of the old authorities. This broadening of the scientific world-view created a new kind of literature, which included eyewitness accounts of nature which observed that it was different from the descriptions in ancient accounts, and it seems that important changes were initiated by such new reports (Taavitsainen forthcoming). New genres emerged to meet the communicative needs of the new discourse community of the Royal Society in the latter half of the seventeenth century. These were characterized by a different way of writing about science and medicine (see below).

5.1. The genre map of early scientific and medical writing

The division of medical texts into the learned tradition, which includes surgical and specialized treatises, and the remedy-book tradition, which includes *materia medica*, is a well-established one (see MEMT design, above). The late medieval period was an important time for the establishing of genre conventions in vernacular writing. Some theoretical treatises were modelled after Latin exemplars and follow the originals closely, but there are also reduced forms and adaptations to more practical uses and applications, and texts with such modifications seem to be even more common than translations. Except perhaps for recipes and charms, discourse forms with their associated text-type features were not yet conventionalized in the vernacular. It is sometimes difficult to tell the layers of texts apart, as references are blurred (Benvenutus), and sentence boundaries are fuzzy (cf. the problems of editing). The picture that emerges from surviving evidence of genres in vernacular medical writing is extremely complicated. A diachronic view can be achieved by comparing late medieval (fifteenth century) and early modern (late seventeenth century) genre maps to one another. Genres change over the course of time, and new genres become added at the highest level, thus changing the positions of other genres. Furthermore, texts from one genre, such as recipes, can occur in several traditions, and genres can vary greatly in terms of their historical significance and transparency, as was the case with patients' case reports. Models of vernacular genres of scientific writing are derived from Latin texts, but English texts may display different features from their original Latin counterparts; furthermore, it is obvious that some of their functions changed in the course of time, and there seems to have been a time lag in the adoption of genres into the vernacular (see below and Taavitsainen 2002, 2004).

Scientific and medical writing in LME and EModE

Learned writing:

Medieval, scholastic	Early Modern; institutional innovations Empirical essays (Royal Society) Book reviews, letters (PT)
Commentaries	~
Compilations →	Textbooks, dialogues
Questions and answers	Handbooks, dialogues

Writing for heterogeneous audiences:

Questions and answers	Textbooks, manuals
Recipes →	
Regimen texts	
Prognostications	Calendar lore, prognostications, medical advice

Figure 1. Scientific and medical writing in the fifteenth and in the late seventeenth century: the baseline.

The foundation of universities from the twelfth century onwards contributed to the formalising of the high-register genres of commentaries, *compilationes* and *questiones*, and the teaching methods of the highest-level institutions of learning were based on these genres (see e.g. Minnis et al. 1988). In the vernacular, these texts lost their original institutional functions and became modified in such a way that their practical sides were enhanced.

Commentaries and *compilationes* were at the heart of the intellectual mainstream of scholasticism. The study of their transfer into the vernacular poses several problems, starting with their identification. Commentaries were used both for research in reconciling ancient authorities and in teaching. The second most formal genre of the learned tradition is that of *compilationes*. This genre was even more important than commentaries for the dissemination of knowledge to the wider public. Originally, the genre had a twofold didactic function: it gave easy access to important passages and authoritative opinions and made them available to readers who could not access the originals (Minnis 1979, 402-3). The techniques of compilation reached a high level of sophistication in medieval encyclopaedic treatises, in which the key doctrines and other *loci communes* are assembled into coherent wholes. These two genres of scholasticism began to converge at the end of the medieval period. It is likely that this development had already begun in Latin writing, with medieval commentaries becoming more similar to compilations, and attitudes changed in the thirteenth and fourteenth centuries (Minnis 1979, 386-7, 413). Genres overlap and merge in vernacular texts, and it is difficult, sometimes impossible, to distinguish between them. In general, vernacular medical texts have a bias towards instruction and practical knowledge, and compilations merge with book-making practices. Thus, the textual histories of texts in these genres are extremely complicated.

The remedy-book tradition with its *materia medica* has more mixed origins, with learned elements intertwining with empirical wisdom and folk beliefs. This tradition can be traced further back in time, to the Old English period, and comprises several different strands with a great deal of intertextuality (see Mäkinen 2006). Recipes in the various traditions of medical writing in Middle English show different degrees of textual standardization: in remedy-books, recipes follow a standardized format, but in academic and surgical texts they are more varied (Taavitsainen 2001c). Recipe collections have recently been approached as "discourse colonies" (Alonso Almeida 2008, 2005, Carroll 2003, Hoey 2001), with the meanings of individual recipes being self-contained and not dependent on the context of adjacent recipes. Recipes are found in multiple contexts, both on their own and embedded within a wide range of texts. It is also intriguing how different genres can be combined in interesting ways (see the example below).

The following passage illustrates the complexity of the genre map of early English medical writing. It comes from Walter Baley's (1529-1592) book on three kinds of peppers from 1588.[15] It shows how a text in any genre can be subject to commentary, using a recipe as an example. The recipe exhibits the typical text-type features of the genre. It begins with the conventionalized imperative form 'take', *recipe*, which gave the genre its name. The list of ingredients includes specific quantities, which is an indication of the learned tradition (in the remedy-book tradition, measurements could be vague, i.e. a handful, a spoonful etc). After the beginning of the recipe, the commentary follows. It discusses the composition of the medicine in the light of ancient authorities, pointing out differences and similarities in their writings. Next, there is a summary of the most important discrepancies, and then the author proceeds to present his own opinion on the matter. The passage fulfils the criteria of a fully fledged commentary (see Minnis 1979: 387).

> Take of the three kinds of pepper +Q .6. +q .ij. ginger, anise seedes, thyme ana. +Q .ij. +q .ij.
> spikenard, amomum ana. +Q .j. +q .j. cassia lignea, asarum, enula camp. dried, persley seeds
> seseli ana. +q .ij. saccharum albissimum q. s.

15 This book (1588) is an important source of evidence for changing thought-styles, as it expresses the growing lack of faith in ancient authorities (Taavitsainen, forthcoming).

This is the same laborious confection set downe by Mesue, of which Galen maketh mention, lib. 4. cap. 11. De sanit. tuend. For Galen in that place doth recite by name euery simple: saue that Galen in stead of Amomum, mentioneth the seeds of ammios. And where our authors do differ in opinion, what is the right Amomum, and all writers do agree, that we haue the right ammi: I adiudge it better in these daies to put in this receit the seeds of ammi, according as Galen prescribeth, than amomum, as Mesue counselleth. So you haue both medicines in vse in our time made of three peppers, and |P_f._B8 are to be sold in the Apothecaries shops, vnder the name of Diatrion pipereon (Baley 1588, f. 16v)

Questiones is the third core genre of scholasticism (Demaitre 1976: 81). It has several layers and traits, all of which develop over time. Originally there were two distinctive trends: the university-based debate and the alternation of questions and answers in pedagogical dialogues, the latter deriving from philosophical writings (Taavitsainen 1999). The academic format cannot be found in the early vernacular texts, but the passage that comes nearest to the academic form uses formal phrases and displays a sequential structure. The text comes from the late fifteenth century, and it is one of the earliest medical incunabula in English:[16]

> But ij=en= questions ys meued in this case: what ys the cause that on dye & the tother not, and in summe tovne the[{r{] deyd men and in another not? The secunde question beth wheter this morbes pestilencial be contagius?
> Atte furste Y do say that this may fortune of ij=en= maner of causes, i. of the parte that cumme of, & of the parte of the pacient. And of the parte that yt cumme of, the influence supercelestiale more directly haue a regarde to thys man or that man, & in that place [{or this place{]. …yt ys to vnderstande …
> And to the ij=de= question, that suche morbes pestilencial be contagius: for cause that the bodyes þat be infected, the humors cumme out of there bodyes & enffect them that be clene of complessyon. Wherffore yt ys goode to voyde … (MEMT; Canutus, Plague treatise pp. 272-273)

The two patterns merge in vernacular literature. The questions are straightforward, without further elaborations, and often the influence

16 STC (2nd ed.), 4592, printed in London by Willelmus de Machlinia in 1485.

of the underlying monological treatise is evident, with long passages being interrupted by occasional blunt questions. The function of disseminating knowledge to wider audiences is evident. The following passage contains metatextual instruction about reading recipes, with explanations of their conventions and a great deal of reader guidance. The extract comes from a fifteenth-century text explicitly targeted at the "poor" for charitable reasons:

> [}Capitulo Sexto}]
> Brother leches haue a queynt maner writyng and hard for to rede in makyng of hir medicynes.
> Brother when ye seeth in bookes of phisique thes writynges R. that is comynly the begynnyng of hir medicines ye shul vnderstond Recipe that is to say take. Ana is as moche for to say as of iche ilich moche. A pound is writen thus li. j. and li. ij and half a pound thus li. di. othur thus li. +s. A quarter of a pound is writen thus qrt. li. An handful is writen thus. M. j. Half [\f. 32v\] an handful is wryten thus M. +s or thus M. di. An vnce is writen thus +o j outher thus j vnce. Half an vnce is writen thus +o +s outher thus +o di. And a quarter of an vnce is writen thus +Q ij. Also a dragme is writen thus +Q j. And half a dragme is writen thus +Q +s outher thus +Q di. A scripulle is written thus +q j. And ye shul vnderstond that thre scripulles maken a dragme and viij dragmes maketh an vnce and xij vncez maketh a pound. And somme men take xvj vnces for a pound as marchauntes don but phisiciens for the more part taketh xij +o for a li. (MEMT; Thesaurus pauperum, f. 32r-v)

In general, there are only a few extant Late Middle English vernacular texts in the dialogue form. The most consistent is perhaps the encyclopaedic treatise *Sidrak and Bokkus*, a compilation of wisdom on all possible topics cast as a teaching dialogue, originally composed in French. The following example contains some practical everyday advice given as an answer to the question of how to remedy a painful condition resulting from swallowing a bone or a "thorne":

> 'Hou shulde a man out of his þrote wynne
> A bone or a þorne stiking þerynne?'
> 'With breed or licour ytake fully
> And swolowe it inward hastily.
> And if it nil for suche a noote
> Falle doun þorgh þe þroote,
> Take of rawe flesshe a morsell

And wiþ a smal þreed binde it wel:
Þat morsel swelowe þou good spede
But in þin honde holde þe þrede;
And whan it is doun, drawe it fast
And it shal þat boone out cast;
And if þe þrede breeke, do it efte
Til þat it be out refte.' (MEMT; *Sidrak and Bokkus* p. 503)

The discourse form of questions and answers becomes increasingly common in the sixteenth century, as it is particularly suitable for instruction. Textbooks of surgery, for example, are often cast in this format in the early modern period. There are several treatises that combine questions and answers with recipes, or genres connected with recipes; for example, William Bullein's *Fever Pestilence*, a social satire and medical handbook, contains long passages including such information. The recipe below is in Latin, demonstrating how common code-switching practices were in this period (see Pahta 2004):

Crispine: … I praie you tell me some verie true experte medicens againste the Pestilence: and I
will write them, and put them in my boke at home. And first of twoo or three siruppes.
Medicus. In deede, for that you coumpte me rude in Englishe: marke what I saie in plaine Latine. A learned man hath with greate modestie after longe studie, written it, I warrant you.
Rec. Syrupi acentositatis citri, ac Syrupi granatis … (EEBO; Bullein, *Fever Pestilence* 1576: 58)

6. Conclusions

The difference between learned and popular layers of writing is reflected in some pertinent features of scholasticism, such as the scale of authorities (see above) and types of prescriptive phrases; the degree of specificity is another indication of the level of a piece of writing (Taavitsainen 2005). It seems possible to trace the transmission of scientific ideas to various audiences through a detailed comparison of the contents and styles of writing of scientific texts. More definitive conclusions would require more studies on the Latin background and more detailed assessments of the features of vernacular writing. Corpora and corpus-linguistic methodology can help us achieve our goals, but the philological approach is important, as it gives us a solid basis with due respect and attention to the form of the language.

The genre map in Figure 1 presents the baseline reality in a simplified form. The above example of a recipe commentary shows that the picture is more complicated and becomes more dynamic when we come to the early modern period. A narrow connecting line from commentaries reaches to recipes, pointing out applications of the above kind in the late sixteenth century. Compilations can also be linked to the lowest layer of texts for heterogeneous audiences, as compilations merged with book-making practices, e.g. in common-place books with collections of useful texts according to the compilers taste and need; this connection applies as early as the fifteenth century. The dynamics of genres provide a fascinating object of study. The work has just begun, and there is a great deal to be done in this field.

References

a. Texts and corpora

MEMT. *Middle English Medical Texts*. Compiled by Taavitsainen, Irma, Päivi Pahta and Martti Mäkinen with software by Raymond Hickey. Amsterdam and Philadelphia: John Benjamins, 2005.

EMEMT. *Early Modern English Medical Texts* 1500-1700. CD-ROM. Taavitsainen, Irma /Päivi Pahta, Turo Hiltunen /Ville Marttila, Martti Mäkinen/Maura Ratia/Carla Suhr/Jukka Tyrkkö (compilers), forthcoming.

Baley, Walter. *A short discourse of the three kindes of peppers in common vse and certaine special medicines made of the same, tending to the preseruation of health.* London: Eliot's Court Press, 1588. EEBO. Early English Books Online. <http://eebo.chadwyck.com/home>.

Bullein William. *A dialogue both pleasant and piety-full, against the fever pestilence.* London: John Kingston, 1564. EEBO. Early English Books Online. <http://eebo.chadwyck.com/home>.

b. Bibliography

Alonso Almeida, Francisco. "All Gathered Together: On the Construction of Scientific and Technical Books in 15th-Century England." *International Journal of English Studies* 5(2). 2005, 1-25.

——. "Middle English medical books as examples of discourse colonies: G.U.L. Hunter 307." In Isabel Moskowich-Spiegel and Begoña Crespo-García (eds.) *Bells Chiming from the Past Cultural and Linguistic Studies on Early English*. Costerus NS 174. Amsterdam/New York, NY: Rodopi, 2008, 55-80.

Atkinson, Dwight. *Scientific Discourse in Sociohistorical Context: The Philosophical Transactions of the Royal Society of London, 1675-1975.* Mahvah, NJ: Laerence Erlbaum, 1999.

Barratt, Alexandra (ed.) *The Knowing of Woman's Kind in Childing: A Middle English Version of Material Derived from the Trotula and Other Sources.* Medieval Women, Texts and Contexts 4. Turnhout: Brepols, 2001, 40-114.

Biber, Douglas. *Variation across Speech and Writing*. Cambridge: Cambridge University Press, 1988.

Burton, T. L. (ed.). Sidrak and Bokkus: A parallel-text edition from Bodleian Library, MS Laud Misc. 559 and British Library, MS Lansdowne 793. I-II. Early English Text Society O.S. 312. Oxford University Press, 1998.

Carroll, Ruth. "Recipes for Laces: An Example of a Middle English Discourse Colony." In Risto Hiltunen and Janne Skaffari (eds.)

Discourse Perspectives on English. Amsterdam and Philadelphia: John Benjamins, 2003, 137-165.

Crisciani, Chiara. "Teachers and learners in scholastic medicine: Some images and metaphors." *History of universities* 15, 2000, 75-101.

Crombie, A. C. *Styles of Scientific Thinking in the European Tradition,* 3 vols. London: Duckworth, 1994.

——. "Commitments and styles of European scientific thinking." *History of Science* 33, 1995, 225-238.

Crossgrove, William. "Introduction." In Crossgrove *et al*. (eds.) 1998, 81-7.

Crossgrove, William, Margaret Schleissner and Linda Ehrsam Voigts (eds.). *Early science and medicine: A journal for the study of science, technology and medicine in the pre-modern period* 3/2, Special issue: *The vernacularization of science, medicine, and technology in Late Medieval Europe*. 1998.

Demaitre, Luke. "Scholasticism in compendia of practical medicine, 1250-1450." *Manuscripta. Special Issue: Science, medicine, and the university: 1200-1550. Essays in honour of Pearl Kibre.* Part I, 1976, 81-95.

Getz, Faye Marie. "Charity, translation, and the language of medical learning in medieval England." *Bulletin of the history of medicine* 64, 1990, 1-17.

Hoey, Michael. *Textual Interaction: An Introduction to Written Discourse Analysis*. London and New York: Routledge, 2001.

Hyland, Ken. *Metadiscourse.* London /New York: Continuum, 2005

Jones, Claire. "Discourse communities and medical texts." In Taavitsainen and Pahta (eds.), 2004, 23-36.

Keiser, George. *A Manual of the Writings in Middle English 1050-1500: Works of Science and Information*. New Haven: The Connecticut Academy of Arts and Sciences, 1998.

Kohnen, Thomas. "Tracing directives through text and time. Towards a methodology of a corpus-based diachronic speech-act analysis." In Andreas H. Jucker and Irma Taavitsainen, *Speech Acts in the History of English*. Amsterdam and Philadelphia: Benjamins, 2008, SIVUT.

____. 2009. "Historical Corpus Pragmatics". In Andreas H. Jucker, Daniel Schreier and Marianne Hundt (eds.) *Corpora: Pragmatics and Discourse. Papers from the 29th International Conference on English Language Research on Computerized Corpora*. Amsterdam: Rodopi. 13-36.

Lass, Roger. "Ut Custodiant Litteras: Editions, Corpora and Witnesshood." In M. Dossena and R. Lass (eds.). *Methods and Data in English Historical Dialectology (Studies in Language and Communication 16)*. Bern: Peter Lang, 2004, 21-48.

McVaugh, Michael R. and Nancy Siraisi (eds). *Renaissance Medical Learning: Evolution of a Tradition,* Osiris 2nd series 6, Philadelphia: University of Pennsylvania Press, 1990.

Minnis, Alastair J. "Late medieval discussions of *compilatio* and the rôle of the *compilator*." *Beiträge zur Geschichte der deutschen Sprache und Literatur* 101, 1979, 385-421.

Minnis, Alastair J., A.B. Scott and David Wallace (eds.). *Medieval literary theory and criticism c.1100-c.1375: The commentary tradition*, Revised edition. Oxford: Clarendon Press, 1988

Mäkinen, Martti. Between Herbals et alia: Intertextuality in Medieval English Herbals. (Diss.) University of Helsinki, 2006. <http://ethesis.helsinki.fi/julkaisut/hum/engla/vk/ makinen/>

Pahta, Päivi. "Code-switching in medieval medical writing." In Taavitsainen and Pahta 2004, 73-99.

Pahta, Päivi and Irma Taavitsainen. "Vernacularisation of scientific and medical writing in its sociohistorical context." In Taavitsainen and Pahta 2004, 1-18.

Ratia, Maura. Argumentative Strategies in the Early Modern Tobacco Controversy. (Diss.) University of Helsinki, forthcoming.

Rhodes, Dennis E. "Provost Argentine of King's and his books." *Transactions of the Cambridge Bibliographical Society* 2, 1956, 205-212.

Robinson, Peter. "Current issues in making digital editions of medieval texts - or, do electronic scholarly editions have a future?" *Digital Medievalist* 1.1., 2005. <http://www. digitalmedievalist.org/article.cfm?RecID=6>.

Siraisi, Nancy. *Medieval and Early Renaissance medicine: An introduction to knowledge and practice*. Chicago and London: University of Chicago Press, 1990.

——. *Medicine and the Italian universities, 1250-1600*. Leiden: Brill, 2001.

Swales, John M. *Genre Analysis: English in Academic and Research Settings*. Cambridge: Cambridge University Press, 1990

Taavitsainen, Irma. "On the evolution of scientific writings from 1375 to 1675: Repertoire of emotive features." In F. Fernandez, M. Fuster and J. Calvo (eds.). *English Historical Linguistics 1992*. Amsterdam/Philadelphia: Benjamins, 1994, 329-42.

____. "Dialogues in Late Medieval and Early Modern English Medical Writing." In Jucker, Andreas H., Gerd Fritz, and Franz Lebsanft (eds). *Historical Dialogue Analysis*. Amsterdam and Philadelphia: Benjamins, 1999, 243-268.

——. "Evidentiality and scientific thought-styles: English medical writing in Late Middle English and Early Modern English." In Maurizio Gotti and Marina Dossena (eds.). *Modality in Specialized Texts: Selected Papers of the 1st CERLIS Conference*. Bern, Berlin, Bruxelles, Frankfurt am Main, New York, Oxford, Wien: Peter Lang, 2001a, 21-52.

——. "Changing conventions of writing: The dynamics of genres, text types, and text traditions." *European Journal of English Studies*, Special issue, ed. by Lilo Moessner, 2001b, 139-150.

——. "Middle English recipes: Genre characteristics, text type features and underlying traditions of writing." *Journal of Historical Pragmatics* 2, 2001c, 85-113.

——. "Historical discourse analysis: Scientific language and changing thought-styles." In Teresa Fanego, Belén Méndez-Naya and Elena Seoane (eds.) *Sounds, Words, Texts, Change. Selected Papers from the Eleventh International Conference on English Historical Linguistics*. Amsterdam and Philadelphia: Benjamins, 2002, 201-226.

——. "Transferring classical discourse conventions into the vernacular." In Irma Taavitsainen and Päivi Pahta (eds). *Medical and scientific writing in Late Medieval English*. Cambridge: Cambridge University Press, 2004, 37-72.

——. "Genres and the appropriation of science: *Loci communes* in English literature in late medieval and early modern periods." In Janne Skaffari, Matti Peikola, Ruth Carroll, Risto Hiltunen and Brita Wårvik (eds.) *Opening Windows on Texts and Discourses of the Past.* Amsterdam and Philadelphia: Benjamins, 2005, 179-196.

——. "Audience Guidance and Learned Medical Writing in Late Medieval English." In Maurizio Gotti and Francoise Salager-Meyer (eds.). *Advances in medical Discourse Analysis: oral and Written Contexts.* Bern: Peter Lang, 2006, 431-456.

——. "Joyful News out of the Newfound World: Medical news reports in Early Modern England." In Andreas H. Jucker (ed.). *News Discourse in a Diachronic Perspective.* Amsterdam and Philadelphia: Benjamins, forthcoming.

Taavitsainen, Irma and Susan Fitzmaurice. "Historical Pragmatics - What Is It and How We Do It?" In Susan Fitzmaurice and Irma Taavitsainen (eds.) *Methods in Historical Pragmatics.* Berlin and New York: Mouton de Gruyter, 2007, 11-36.

Taavitsainen, Irma and Päivi Pahta. "Scientific Thought-styles in discourse structure: Changing patterns in a historical perspective." In Wårvik et al. (eds.). *Organization in Discourse: Proceedings from the Turku Conference.* Anglicana Turkuensia 14, University of Turku, 1995, 519-529.

——. "The Corpus of Early Medical Writing: Linguistic Variation and Prescriptive Collocations in Scholastic Style." In Terttu Nevalainen, and Leena Kahalas-Tarkka (eds.). *To Explain the Present: Studies in the Changing English Language in Honour of Matti Rissanen,* Mémoires de la Société Néophilologique de Helsinki, 1997, 209-25.

——. "Vernacularisation of Medical Writing in English: A Corpus-based Study of Scholasticism". In Crossgrove et al. 1998, 157-185.

____ (eds). *Medical and Scientific Writing in Late Medieval English.* Cambridge: Cambridge University Press, 2004.

Tavormina, M. Teresa (ed.). Sex, Aging, & Death in a Medieval Medical Compendium: *Trinity College Cambridge MS R.14.5,*

Its Texts Language, and scribe. Tempe: Arizona Center for Medieval and Renaissance Studies, 2006.

Voigts, Linda Ehrsam and Patricia Deery Kurtz (eds.). *Scientific and medical writings in Old and Middle English: An electronic reference*. (eVK) CD-ROM. Ann Arbor: University of Michigan Press, 1998.

Voigts, Linda E. et al. Electronic Thorndike Kibre Project (eTK): L. Thorndike and P. Kibre Catalog of *Medieval scientific writings in Latin* (1963, 1965, 1968), forthcoming.

Webster, Charles (ed.). *Health, Medicine and Mortality in the Sixteenth Century*. Cambridge: Cambridge University Press, 1979.

Wogan-Browne, Jocelyn, Nicholas Watson, Andrew Taylor and Ruth Evans (eds.). *The idea of the vernacular: An anthology of Middle English literary theory, 1280-1520*. Exeter: University of Exeter Press, 1999.

Index

Notes on Contributors

FRANCISCO ALONSO-ALMEIDA is Senior Lecturer of English Historical Linguistics at the University of Las Palmas de Gran Canaria. His research interests include aspects of historical linguistics and manuscript studies, although his major research area is in the field of medical manuscripts and texts from a pragmatic and discourse perspective. Currently he is leading a project, which aims at the construction of a corpus of English recipes from 1375 to 1850, the *Corpus of Early English Recipes* (CoER). He has published extensively on recipes and their linguistic configuration in leading journals of the area. He also works on PDE scientific texts to study aspects related to modality. He is team leader of the project *Evidentiality of a Multisciplinary Corpus of English Scientific Papers*, funded by the Spanish Ministry of Science and Innovation. He also collaborates in a project dealing with the remote identification of users through handwriting and biometric parameters (project funded by Telefónica).

ROSARIO CABALLERO is Associate Professor in English Language at the University of Castilla-La Mancha. Her main research areas are Cognitive Linguistics, and Genre and Discourse Analysis. Over the past years she has been working on issues related to metaphor in various genres and has given several invited talks on the topic (universities of Florence, Växjo, Zaragoza, Logroño). She has published several articles on these topics in international journals such as *English for Specific Purposes* ("How to talk shop through metaphor: Bringing metaphor research to the ESP classroom"), *Applied Linguistics* ("Metaphor and genre: The presence and role of metaphor in the building review"), *Annual Review of Cognitive Linguistics* ("Talking about space: Image metaphor in architectural discourse") and *Journal of Pragmatics* ("Manner*-of-motion verbs in wine description"), as well as the volume *Re-viewing space. Figurative language in architects' assessment of built space* (2006).

GRAHAM CAIE is a Professor of English Language at the University of Glasgow. His research and teaching interests include Old and Middle English language and literature, the history of the English language, Scots language, interdisciplinary medieval studies, medieval drama, literacy, codicology and editing. He initiated the Middle English Manuscript Digitisation project, which aims at preparing digitised images, accompanied by transcriptions, of Glasgow University Library's Hunterian Collection of Middle English manuscripts. His publications include *The Judgement Day Theme in Old English Poetry* (1976), *A Bibliography of MS Junius XI* (1979), *The Old English Judgement Day II: A Critical Edition with editions of Bede's De die iudicii and the Hatton 113 Homily Be Domes Dæge* (2000) and books on Chaucer and Beowulf, as well as numerous book chapters and articles on Old and Middle English, glosses, language and cultural identity, medieval art and drama, and Arthurian literature. In progress is an edition of *The Man of Law's Tale* for the Variorum Chaucer Project and a book with Denis Renevey entitled *Medieval Texts in Manuscript Context.*

JAVIER CALLE-MARTÍN is Senior Lecturer of English Historical Linguistics in the Department of English at the University of Málaga (Spain). His research interests range from Historical Linguistics to Palaeography, Codicology, Manuscript Transmission and Authorship Attribution Studies. He has published in specialised journals such as *Neuphilologische Mitteilungen* (2004), *Folia Linguistica Historica* (2005), *Literary and Linguistic Computing* (2007), *Studia Neophilologica* (2007) and *Review of English Studies* (2008). He is currently the leading researcher of a project funded by the Autonomous Government of Andalusia (P07-HUM02609) aiming at (a) the electronic editing of late Middle English *Fachprosa* and (b) the compilation of an annotated corpus of mediaeval science in the vernacular.

MARK CHAMBERS is a Research Assistant on the Lexis of Cloth and Clothing Project based at the University of Westminster, London, UK. He holds a PhD in English literature from Trinity College Dublin (IRE). He has lectured on medieval English language and literature at the University of Durham and at Trinity College Dublin and has published articles on late-medieval dramatic allegory and costuming. He is currently conducting research on fashion and sumptuary legislation in the 14th through the 16th centuries and on the lexicology of medieval fashion. He is also preparing work on violence in the late-medieval drama and on the history of the tutelary good and bad angels in Western literature.

JAVIER E. DÍAZ-VERA is Senior Lecturer of English Historical Linguistics in the Department of Modern Philology at the University of Castilla-La Mancha (Spain). His research interests include Historical Linguistics, Sociolinguistics and Cognitive Linguistics. He has published a wide variety of papers on Old and Middle English semantics and lexicology, including "Lexical and non-lexical linguistic variation in the vocabulary of Old English" (2003), "Image schemata and light: A study in diachronic lexical domains in English" (2004), "Calculating indexes of lexical convergence and divergence through time and space" (2006) and "Metaphors we learnt by: Cultural traditions and metaphorical patterns in the Old English vocabulary of KNOWLEDGE" (2007).

MARÍA LAURA ESTEBAN-SEGURA is a Lecturer of English at the University of Murcia. She graduated in English Language and Literature (BA), Translation and Interpreting (BA) and Specialized Translation and Interpreting (MA) from the University of Málaga, and in English Language and English Linguistics (MLitt) from the University of Glasgow. In 2008 she was awarded a PhD in English Linguistics (University of Málaga). She is author of a number of articles on the medical work known as the Middle English Gilbertus Anglicus. Her research interests include English historical linguistics, manuscripts studies and translation.

TERESA MARQUÉS-AGUADO is a Lecturer in the Department of English at the University of Murcia, where she teaches English Language. She holds a BA in English Language (University of Málaga), a MA in Specialized Translation and Interpreting (University of Málaga), and a MLitt in English Language and English Linguistics

(University of Glasgow). In 2008, she completed her PhD Thesis on the edition and philological study of the *Antidotary* contained in Glasgow University Library, MS Hunter 513. Her research interests include palaeography, codicology, manuscript studies and historical linguistics. In connection with them, she is currently involved in two research projects (FFI2008-02336/FILO "Grupo Consolidado", and Proyecto de Excelencia P07-HUM-02609, granted by the Spanish Ministry of Science and Innovation and the regional government of the Junta de Andalucía, respectively) consisting in the electronic edition of unedited late Middle English texts, which will in turn provide the input for an annotated corpus of late Middle English scientific prose.

ANTONIO MIRANDA-GARCÍA has been an associate Professor (CEU) at the University of Málaga since 1994, where hes has taught *History of the English Language* (an introductory course for undergraduates, Faculty of Arts), *Technical English* (undergraduate, School of Engineering), and a post-graduate course entitled *Electronic Resources for Linguistic Research.* He has also coordinated the research group HUM 497 funded by Junta of Andalucia since 1996 and is the research leader of two projects funded by the Ministry of Science and Technology (BFF 2001/1835 and HUM2004/01075FILO). His research interests are Vocabulary Acquisition, ESP, Historical Linguistics, and Authorship Attribution studies.

DAVID MORENO-OLALLA (BA, MA, PhD) is a Lecturer at the Department of English, French and German Philology of the University of Málaga. Most of his published work deals with Old and Middle English *Fachliteratur* –in especial, Medicine and Botany–, MSS studies (mainly Stemmatics), Dialectology, and Historical Corpus Linguistics. He also pursues a strong interest in the historical aspects of Phonology and Morphology, particularly with reference to Germanic and early OE, and Comparative Linguistics. He is currently finishing a critical and annotated edition of a fourteenth-century unpublished Middle English translation of Macer Floridus' *De Virtutis Herbarum.*

NADIA OBEGI GALLARDO is a part-time lecturer in the Department of English at the University of Málaga (Spain) where she teaches English for Specific Purposes in Audiovisual Communication / Media Studies. Her research interests are Codicology, Palaeography and English Historical Linguistics. She is now finishing her PhD Dissertation on the critical edition of G.U.L. MS Hunter 95 keeping a Middle English version of a *Book of Operation*.

ELENA QUINTANA-TOLEDO graduated in English Philology from the University of Las Palmas de Gran Canaria where she is currently pursuing a PhD on Historical Pragmatics. She holds a Research Fellowship from the same university. She is a member of the research project *Corpus of Early English Recipes* (CoER) and collaborates in *Evidentiality in a Multidisciplinary Corpus of Research Papers in English*, a project funded by the Spanish Ministry of Science and Innovation. Her research interests include English language, the pragmatic and discursive analysis of medical texts, and corpora construction.

IRMA TAAVITSAINEN is a Professor of English Philology at the University of Helsinki. Her research interests cover historical pragmatics and corpus linguistics, historical discourse analysis, genre and text type studies, diachronic studies on register variation, and the position of English in Finland. She has co-authored the volume *Medical and Scientific Writing in Late Medieval English* (eds Irma Taavitsainen and Päivi Pahta, 2004), based on the corpus of *Middle English Medical Texts1375-1500* (MEMT, compiled by Irma Taavitsainen, Päivi Pahta and Martti Mäkinen, 2005). Her new book will be based on the corpus of *Early Modern English Medical Texts 1500-1700* (forthcoming in 2009).

Linguistic Insights

Studies in Language and Communication

This series aims to promote specialist language studies in the fields of linguistic theory and applied linguistics, by publishing volumes that focus on specific aspects of language use in one or several languages and provide valuable insights into language and communication research. A cross-disciplinary approach is favoured and most European languages are accepted.

The series includes two types of books:

- **Monographs** – featuring in-depth studies on special aspects of language theory, language analysis or language teaching.
- **Collected papers** – assembling papers from workshops, conferences or symposia.

Each volume of the series is subjected to a double peer-reviewing process.

Vol. 1 Maurizio Gotti & Marina Dossena (eds)
Modality in Specialized Texts. Selected Papers of the 1st CERLIS Conference.
421 pages. 2001. ISBN 3-906767-10-8. US-ISBN 0-8204-5340-4

Vol. 2 Giuseppina Cortese & Philip Riley (eds)
Domain-specific English. Textual Practices across Communities and Classrooms.
420 pages. 2002. ISBN 3-906768-98-8. US-ISBN 0-8204-5884-8

Vol. 3 Maurizio Gotti, Dorothee Heller & Marina Dossena (eds)
Conflict and Negotiation in Specialized Texts. Selected Papers of the 2nd CERLIS Conference.
470 pages. 2002. ISBN 3-906769-12-7. US-ISBN 0-8204-5887-2

Editorial address:

Prof. Maurizio Gotti — Università di Bergamo, Facoltà di Lingue e Letterature Straniere,
Via Salvecchio 19, 24129 Bergamo, Italy
Fax: 0039 035 2052789, E-Mail: m.gotti@unibg.it

Vol. 4 Maurizio Gotti, Marina Dossena, Richard Dury, Roberta Facchinetti & Maria Lima
Variation in Central Modals. A Repertoire of Forms and Types of Usage in Middle English and Early Modern English.
364 pages. 2002. ISBN 3-906769-84-4. US-ISBN 0-8204-5898-8

Vol. 5 Stefania Nuccorini (ed.)
Phrases and Phraseology. Data and Descriptions.
187 pages. 2002. ISBN 3-906770-08-7. US-ISBN 0-8204-5933-X

Vol. 6 Vijay Bhatia, Christopher N. Candlin & Maurizio Gotti (eds)
Legal Discourse in Multilingual and Multicultural Contexts. Arbitration Texts in Europe.
385 pages. 2003. ISBN 3-906770-85-0. US-ISBN 0-8204-6254-3

Vol. 7 Marina Dossena & Charles Jones (eds)
Insights into Late Modern English. 2nd edition.
378 pages. 2003, 2007.
ISBN 978-3-03911-257-9. US-ISBN 978-0-8204-8927-8

Vol. 8 Maurizio Gotti
Specialized Discourse. Linguistic Features and Changing Conventions.
351 pages. 2003, 2005.
ISBN 3-03910-606-6. US-ISBN 0-8204-7000-7

Vol. 9 Alan Partington, John Morley & Louann Haarman (eds)
Corpora and Discourse.
420 pages. 2004. ISBN 3-03910-026-2. US-ISBN 0-8204-6262-4

Vol. 10 Martina Möllering
The Acquisition of German Modal Particles. A Corpus-Based Approach.
290 pages. 2004. ISBN 3-03910-043-2. US-ISBN 0-8204-6273-X

Vol. 11 David Hart (ed.)
English Modality in Context. Diachronic Perspectives.
261 pages. 2003. ISBN 3-03910-046-7. US-ISBN 0-8204-6852-5

Vol. 12 Wendy Swanson
Modes of Co-reference as an Indicator of Genre.
430 pages. 2003. ISBN 3-03910-052-1. US-ISBN 0-8204-6855-X

Vol. 13 Gina Poncini
Discursive Strategies in Multicultural Business Meetings.
2nd edition. 338 pages. 2004, 2007.
ISBN 978-3-03911-296-8. US-ISBN 978-0-8204-8937-7

Vol. 14 Christopher N. Candlin & Maurizio Gotti (eds)
Intercultural Aspects of Specialized Communication.
2nd edition. 369 pages. 2004, 2007.
ISBN 978-3-03911-258-6. US-ISBN 978-0-8204-8926-1

Vol. 15 Gabriella Del Lungo Camiciotti & Elena Tognini Bonelli (eds)
Academic Discourse. New Insights into Evaluation.
234 pages. 2004. ISBN 3-03910-353-9. US-ISBN 0-8204-7016-3

Vol. 16 Marina Dossena & Roger Lass (eds)
Methods and Data in English Historical Dialectology.
405 pages. 2004. ISBN 3-03910-362-8. US-ISBN 0-8204-7018-X

Vol. 17 Judy Noguchi
The Science Review Article. An Opportune Genre in the Construction of Science.
274 pages. 2006. ISBN 3-03910-426-8. US-ISBN 0-8204-7034-1

Vol. 18 Giuseppina Cortese & Anna Duszak (eds)
Identity, Community, Discourse. English in Intercultural Settings.
495 pages. 2005. ISBN 3-03910-632-5. US-ISBN 0-8204-7163-1

Vol. 19 Anna Trosborg & Poul Erik Flyvholm Jørgensen (eds)
Business Discourse. Texts and Contexts.
250 pages. 2005. ISBN 3-03910-606-6. US-ISBN 0-8204-7000-7

Vol. 20 Christopher Williams
Tradition and Change in Legal English. Verbal Constructions in Prescriptive Texts.
2nd revised edition. 216 pages. 2005, 2007. ISBN 978-3-03911-444-3.

Vol. 21 Katarzyna Dziubalska-Kolaczyk & Joanna Przedlacka (eds)
English Pronunciation Models: A Changing Scene.
2nd edition. 476 pages. 2005, 2008. ISBN 978-3-03911-682-9.

Vol. 22 Christián Abello-Contesse, Rubén Chacón-Beltrán, M. Dolores López-Jiménez & M. Mar Torreblanca-López (eds)
Age in L2 Acquisition and Teaching.
214 pages. 2006. ISBN 3-03910-668-6. US-ISBN 0-8204-7174-7

Vol. 23 Vijay K. Bhatia, Maurizio Gotti, Jan Engberg & Dorothee Heller (eds)
Vagueness in Normative Texts.
474 pages. 2005. ISBN 3-03910-653-8. US-ISBN 0-8204-7169-0

Vol. 24 Paul Gillaerts & Maurizio Gotti (eds)
Genre Variation in Business Letters. 2nd printing.
407 pages. 2008. ISBN 978-3-03911-681-2.

Vol. 25 Ana María Hornero, María José Luzón & Silvia Murillo (eds)
Corpus Linguistics. Applications for the Study of English.
2nd printing. 526 pages. 2006, 2008. ISBN 978-3-03911-726-0

Vol. 26 J. Lachlan Mackenzie & María de los Ángeles Gómez-González (eds)
Studies in Functional Discourse Grammar.
259 pages. 2005. ISBN 3-03910-696-1 / US-ISBN 0-8204-7558-0

Vol. 27 Debbie G. E. Ho
Classroom Talk. Exploring the Sociocultural Structure of Formal ESL Learning.
2nd edition. 254 pages. 2006, 2007. ISBN 978-3-03911-434-4

Vol. 28 Javier Pérez-Guerra, Dolores González-Álvarez, Jorge L. Bueno-Alonso & Esperanza Rama-Martínez (eds)
'Of Varying Language and Opposing Creed': New Insights into Late Modern English.
455 pages. 2007. ISBN 978-3-03910-788-9

Vol. 29 Francesca Bargiela-Chiappini & Maurizio Gotti (eds)
Asian Business Discourse(s).
350 pages. 2005. ISBN 3-03910-804-2 / US-ISBN 0-8204-7574-2

Vol. 30 Nicholas Brownlees (ed.)
News Discourse in Early Modern Britain. Selected Papers of CHINED 2004.
300 pages. 2006. ISBN 3-03910-805-0 / US-ISBN 0-8204-8025-8

Vol. 31 Roberta Facchinetti & Matti Rissanen (eds)
Corpus-based Studies of Diachronic English.
300 pages. 2006. ISBN 3-03910-851-4 / US-ISBN 0-8204-8040-1

Vol. 32 Marina Dossena & Susan M. Fitzmaurice (eds)
Business and Official Correspondence: Historical Investigations.
209 pages. 2006. ISBN 3-03910-880-8 / US-ISBN 0-8204-8352-4

Vol. 33 Giuliana Garzone & Srikant Sarangi (eds)
Discourse, Ideology and Specialized Communication.
494 pages. 2007. ISBN 978-3-03910-888-6

Vol. 34 Giuliana Garzone & Cornelia Ilie (eds)
The Use of English in Institutional and Business Settings.
An Intercultural Perspective.
372 pages. 2007. ISBN 978-3-03910-889-3

Vol. 35 Vijay K. Bhatia & Maurizio Gotti (eds)
Explorations in Specialized Genres.
316 pages. 2006. ISBN 3-03910-995-2 / US-ISBN 0-8204-8372-9

Vol. 36 Heribert Picht (ed.)
Modern Approaches to Terminological Theories and Applications.
432 pages. 2006. ISBN 3-03911-156-6 / US-ISBN 0-8204-8380-X

Vol. 37 Anne Wagner & Sophie Cacciaguidi-Fahy (eds)
Legal Language and the Search for Clarity / Le langage juridique et la quête de clarté.
Practice and Tools / Pratiques et instruments.
487 pages. 2006. ISBN 3-03911-169-8 / US-ISBN 0-8388-5

Vol. 38 Juan Carlos Palmer-Silveira, Miguel F. Ruiz-Garrido &
Inmaculada Fortanet-Gómez (eds)
Intercultural and International Business Communication:
Theory, Research and Teaching.
2nd edition. 343 pages. 2006, 2008. ISBN 978-3-03911-680-5

Vol. 39 Christiane Dalton-Puffer, Dieter Kastovsky, Nikolaus Ritt &
Herbert Schendl (eds)
Syntax, Style and Grammatical Norms: English from 1500–2000.
250 pages. 2006. ISBN 3-03911-181-7 / US-ISBN 0-8204-8394-X

Vol. 40 Marina Dossena & Irma Taavitsainen (eds)
Diachronic Perspectives on Domain-Specific English.
280 pages. 2006. ISBN 3-03910-176-0 / US-ISBN 0-8204-8391-5

Vol. 41 John Flowerdew & Maurizio Gotti (eds)
Studies in Specialized Discourse.
293 pages. 2006. ISBN 3-03911-178-7

Vol. 42 Ken Hyland & Marina Bondi (eds)
Academic Discourse Across Disciplines.
320 pages. 2006. ISBN 3-03911-183-3 / US-ISBN 0-8204-8396-6

Vol. 43 Paul Gillaerts & Philip Shaw (eds)
The Map and the Landscape: Norms and Practices in Genre.
256 pages. 2006. ISBN 3-03911-182-5 / US-ISBN 0-8204-8395-4

Vol. 44 Maurizio Gotti & Davide Giannoni (eds)
New Trends in Specialized Discourse Analysis.
301 pages. 2006. ISBN 3-03911-184-1 / US-ISBN 0-8204-8381-8

Vol. 45 Maurizio Gotti & Françoise Salager-Meyer (eds)
Advances in Medical Discourse Analysis: Oral and Written Contexts.
492 pages. 2006. ISBN 3-03911-185-X / US-ISBN 0-8204-8382-6

Vol. 46 Maurizio Gotti & Susan Šarcević (eds)
Insights into Specialized Translation.
396 pages. 2006. ISBN 3-03911-186-8 / US-ISBN 0-8204-8383-4

Vol. 47 Khurshid Ahmad & Margaret Rogers (eds)
Evidence-based LSP: Translation, Text and Terminology.
584 pages. 2007. ISBN 978-3-03911-187-9

Vol. 48 Hao Sun & Dániel Z. Kádár (eds)
It's the Dragon's Turn: Chinese Institutional Discourses.
262 pages. 2008. ISBN 978-3-03911-175-6

Vol. 49 Cristina Suárez-Gómez
Relativization in Early English (950-1250): the Position of Relative Clauses.
149 pages. 2006. ISBN 3-03911-203-1 / US-ISBN 0-8204-8904-2

Vol. 50 Maria Vittoria Calvi & Luisa Chierichetti (eds)
Nuevas tendencias en el discurso de especialidad.
319 pages. 2006. ISBN 978-3-03911-261-6

Vol. 51 Mari Carmen Campoy & María José Luzón (eds)
Spoken Corpora in Applied Linguistics.
274 pages. 2008. ISBN 978-3-03911-275-3

Vol. 52 Konrad Ehlich & Dorothee Heller (Hrsg.)
Die Wissenschaft und ihre Sprachen.
323 pages. 2006. ISBN 978-3-03911-272-2

Vol. 53 Jingyu Zhang
The Semantic Salience Hierarchy Model: The L2 Acquisition of Psych Predicates
273 pages. 2007. ISBN 978-3-03911-300-2

Vol. 54 Norman Fairclough, Giuseppina Cortese & Patrizia Ardizzone (eds)
Discourse and Contemporary Social Change.
555 pages. 2007. ISBN 978-3-03911-276-0

55 Jan Engberg, Marianne Grove Ditlevsen, Peter Kastberg & Martin Stegu (eds)
New Directions in LSP Teaching.
331 pages. 2007. ISBN 978-3-03911-433-7

Vol. 56 Dorothee Heller & Konrad Ehlich (Hrsg.)
Studien zur Rechtskommunikation.
322 pages. 2007. ISBN 978-3-03911-436-8

Vol. 57 Forthcoming.

Vol. 58 Carmen Frehner
Email – SMS – MMS
294 pages. 2008. ISBN 978-3-03911-451-1

Vol. 59 Isabel Balteiro
The Directionality of Conversion in English: A Dia-Synchronic Study.
276 pages. 2007. ISBN 978-3-03911-241-8

Vol. 60 Maria Milagros Del Saz Rubio
English Discourse Markers of Reformulation.
237 pages. 2007. ISBN 978-3-03911-196-1

Vol. 61 Sally Burgess & Pedro Martín-Martín (eds)
English as an Additional Language in Research Publication and Communication.
259 pages. 2008. ISBN 978-3-03911-462-7

Vol. 62 Sandrine Onillon
Pratiques et représentations de l'écrit.
458 pages. 2008. ISBN 978-3-03911-464-1

Vol. 63 Hugo Bowles & Paul Seedhouse (eds)
Conversation Analysis and Language for Specific Purposes.
2nd edition. 337 pages. 2007, 2009. ISBN 978-3-0343-0045-2

Vol. 64 Vijay K. Bhatia, Christopher N. Candlin & Paola Evangelisti Allori (eds)
Language, Culture and the Law.
The Formulation of Legal Concepts across Systems and Cultures.
342 pages. 2008. ISBN 978-3-03911-470-2

Vol. 65 Jonathan Culpeper & Dániel Z. Kádár (eds)
Historical (Im)politeness.
300 pages. 2010. ISBN 978-3-03911-496-2

Vol. 66 Linda Lombardo (ed.)
Using Corpora to Learn about Language and Discourse.
237 pages. 2009. ISBN 978-3-03911-522-8

Vol. 67 Natsumi Wakamoto
Extroversion/Introversion in Foreign Language Learning.
Interactions with Learner Strategy Use.
159 pages. 2009. ISBN 978-3-03911-596-9

Vol. 68 Eva Alcón-Soler (ed.)
Learning How to Request in an Instructed Language Learning Context.
260 pages. 2008. ISBN 978-3-03911-601-0

Vol. 69 Domenico Pezzini
The Translation of Religious Texts in the Middle Ages.
428 pages. 2008. ISBN 978-3-03911-600-3

Vol. 70 Tomoko Tode
Effects of Frequency in Classroom Second Language Learning.
Quasi-experiment and stimulated-recall analysis.
195 pages. 2008. ISBN 978-3-03911-602-7

Vol. 71 Egor Tsedryk
Fusion symétrique et alternances ditransitives.
211 pages. 2009. ISBN 978-3-03911-609-6

Vol. 72 Cynthia J. Kellett Bidoli & Elana Ochse (eds)
English in International Deaf Communication.
444 pages. 2008. ISBN 978-3-03911-610-2

Vol. 73 Joan C. Beal, Carmela Nocera & Massimo Sturiale (eds)
Perspectives on Prescriptivism.
269 pages. 2008. ISBN 978-3-03911-632-4

Vol. 74 Carol Taylor Torsello, Katherine Ackerley & Erik Castello (eds)
Corpora for University Language Teachers.
308 pages. 2008. ISBN 978-3-03911-639-3

Vol. 75 María Luisa Pérez Cañado (ed.)
English Language Teaching in the European Credit Transfer System: Facing the Challenge.
251 pages. 2009. ISBN 978-3-03911-654-6

Vol. 76 Marina Dossena & Ingrid Tieken-Boon van Ostade (eds)
Studies in Late Modern English Correspondence. Methodology and Data.
291 pages. 2008. ISBN 978-3-03911-658-4

Vol. 77 Ingrid Tieken-Boon van Ostade & Wim van der Wurff (eds)
Current Issues in Late Modern English.
436 pages. 2009. ISBN 978-3-03911-660-7

Vol. 78 Marta Navarro Coy (ed.)
Practical Approaches to Foreign Language Teaching and Learning.
297 pages. 2009. ISBN 978-3-03911-661-4

Vol. 79 Qing Ma
Second Language Vocabulary Acquisition.
333 pages. 2009. ISBN 978-3-03911-666-9

Vol. 80 Martin Solly, Michelangelo Conoscenti & Sandra Campagna (eds)
Verbal/Visual Narrative Texts in Higher Education.
384 pages. 2008. ISBN 978-3-03911-672-0

Vol. 81 Meiko Matsumoto
From Simple Verbs to Periphrastic Expressions:
The Historical Development of Composite Predicates, Phrasal Verbs, and Related Constructions in English.
235 pages. 2008. ISBN 978-3-03911-675-1

82 Melinda Dooly
Doing Diversity. Teachers' Construction of Their Classroom Reality.
180 pages. 2009. ISBN 978-3-03911-687-4

Vol. 83 Victoria Guillén-Nieto, Carmen Marimón-Llorca & Chelo Vargas-Sierra (eds)
Intercultural Business Communication and
Simulation and Gaming Methodology.
392 pages. 2009. ISBN 978-3-03911-688-1

Vol. 84 Maria Grazia Guido
English as a Lingua Franca in Cross-cultural Immigration Domains.
285 pages. 2008. ISBN 978-3-03911-689-8

Vol. 85 Erik Castello
Text Complexity and Reading Comprehension Tests.
352 pages. 2008. ISBN 978-3-03911-717-8

Vol. 86 Forthcoming.

Vol. 87 Carmen Navarro, Rosa Mª Rodríguez Abella, Francesca Dalle Pezze
& Renzo Miotti (eds)
La comunicación especializada.
355 pages. 2008. ISBN 978-3-03911-733-8

Vol. 88 Kiriko Sato
The Development from Case-Forms to Prepositional Constructions
in Old English Prose.
231 pages. 2009. ISBN 978-3-03911-763-5

Vol. 89 Dorothee Heller (Hrsg.)
Formulierungsmuster in deutscher und italienischer Fachkommunikation.
Intra- und interlinguale Perspektiven.
315 pages. 2008. ISBN 978-3-03911-778-9

Vol. 90 Henning Bergenholtz, Sandro Nielsen & Sven Tarp (eds)
Lexicography at a Crossroads. Dictionaries and Encyclopedias Today,
Lexicographical Tools Tomorrow.
372 pages. 2009. ISBN 978-3-03911-799-4

Vol. 91 Manouchehr Moshtagh Khorasani
The Development of Controversies: From the Early Modern Period
to Online Discussion Forums.
317 pages. 2009. ISBN 978-3-3911-711-6

Vol. 92 Maria Luisa Carrió-Pastor (ed.)
Content and Language Integrated Learning: Cultural Diversity.
178 pages. 2009. ISBN 978-3-3911-818-2

Vol. 93-96 Forthcoming.

Vol. 97 Sylvie Hancil (ed.)
The Role of Prosody in Affective Speech.
403 pages. 2009. ISBN 978-3-03911-696-6

Vol. 98 Marina Dossena & Roger Lass (eds)
Studies in English and European Historical Dialectology.
257 pages. 2009. ISBN 978-3-0343-0024-7

Vol. 99 Forthcoming.

Vol. 100 Maurizio Gotti (ed.)
Commonality and Individuality in Academic Discourse.
398 pages. 2009. ISBN 978-3-0343-0023-0

Vol. 101 Javier E. Díaz Vera & Rosario Caballero (eds)
Textual Healing: Studies in Medieval English Medical, Scientific and Technical Texts.
213 pages. 2009. ISBN 978-3-03911-822-9

Vol. 102 Nuria Edo Marzá
The Specialised Lexicographical Approach: A Step further in Dictionary-making.
316 pages. 2009. ISBN 978-3-0343-0043-8

Vol. 103 Carlos Prado-Alonso, Lidia Gómez-García, Iria Pastor-Gómez & David Tizón-Couto (eds)
New Trends and Methodologies in Applied English Language Research. Diachronic, Diatopic and Contrastive Studies.
348 pages. 2009. ISBN 978-3-0343-0046-9

Vol. 104-108 Forthcoming.

Vol. 109 Ángeles Linde López & Rosalía Crespo Jiménez (eds)
Professional English in the European context: The EHEA challenge.
374 pages. 2010. ISBN 978-3-0343-0088-9

Vol. 110 Rosalía Rodríguez-Vázquez
The Rhythm of Speech, Verse and Vocal Music: A New Theory.
394 pages. 2010. ISBN 978-3-0343-0309-5

Vol. 111 Anastasios Tsangalidis & Roberta Facchinetti (eds)
Studies on English Modality. In Honour of Frank Palmer.
392 pages. 2009. ISBN 978-3-0343-0310-1

Vol. 112-114 Forthcoming.

Vol. 115 Giuliana Garzone & Paola Catenaccio (eds)
Identities across Media and Modes: Discursive Perspectives.
379 pages. 2009. ISBN 978-3-0343-0386-6

Vol. 116 Elena Landone
Los marcadores del discurso y cortesía verbal en español.
390 pages. 2010. ISBN 978-3-0343-0413-9